T0263838

# Molecular Tools and Infectious Disease Epidemiology

# Molecular Tools and Infectious Disease Epidemiology

**Betsy Foxman**

AMSTERDAM • BOSTON • HEIDELBERG • LONDON • NEW YORK • OXFORD
PARIS • SAN DIEGO • SAN FRANCISCO • SINGAPORE • SYDNEY • TOKYO

Academic Press is an imprint of Elsevier

Academic Press is an imprint of Elsevier
30 Corporate Drive, Suite 400, Burlington, MA 01803, USA
525 B Street, Suite 1800, San Diego, California 92101-4495, USA
84 Theobald's Road, London WC1X 8RR, UK

Copyright © 2012, Elsevier Inc. All rights reserved

No part of this publication may be reproduced or transmitted in any form or by any means, electronic or mechanical, including photocopying, recording, or any information storage and retrieval system, without permission in writing from the publisher. Details on how to seek permission, further information about the Publisher's permissions policies and our arrangements with organizations such as the Copyright Clearance Center and the Copyright Licensing Agency, can be found at our website: www.elsevier.com/permissions.

This book and the individual contributions contained in it are protected under copyright by the Publisher (other than as may be noted herein).

**Notices**
Knowledge and best practice in this field are constantly changing. As new research and experience broaden our understanding, changes in research methods, professional practices, or medical treatment may become necessary.

Practitioners and researchers must always rely on their own experience and knowledge in evaluating and using any information, methods, compounds, or experiments described herein. In using such information or methods they should be mindful of their own safety and the safety of others, including parties for whom they have a professional responsibility.

To the fullest extent of the law, neither the Publisher nor the authors, contributors, or editors, assume any liability for any injury and/or damage to persons or property as a matter of products liability, negligence or otherwise, or from any use or operation of any methods, products, instructions, or ideas contained in the material herein.

**Library of Congress Cataloging-in-Publication Data**
Foxman, Betsy, 1955-
Molecular tools and infectious disease epidemiology / Betsy Foxman.
  p. ; cm.
Includes bibliographical references and index.
ISBN 978-0-12-374133-2
1. Communicable diseases—Epidemiology. 2. Molecular epidemiology. I. Title.
[DNLM: 1. Communicable Diseases—epidemiology. 2. Communicable Diseases—genetics.
3. Molecular Epidemiology. WC 100].
RA643.F69 2012
614.4—dc22                                                          2010042367

**British Library Cataloguing-in-Publication Data**
A catalogue record for this book is available from the British Library.

For information on all Academic Press publications
visit our web site at www.elsevierdirect.com

Working together to grow
libraries in developing countries

www.elsevier.com | www.bookaid.org | www.sabre.org

ELSEVIER    BOOK AID International    Sabre Foundation

# CONTENTS

## WHY A TEXTBOOK ON MOLECULAR EPIDEMIOLOGY OF INFECTIOUS DISEASE?

As a Professor of Epidemiology in the Hospital and Molecular Epidemiology masters' program at the University of Michigan, I frequently meet with potential students. Because our program includes laboratory training, many of these students have undergraduate degrees in biology or microbiology, and extensive laboratory experience. Why are they considering a degree in epidemiology? The most common answer is that although they enjoy lab work, they want to be in a position to see how their work at the bench "makes a difference." Making a difference in human health is a core value of epidemiology, the science that uses field, laboratory, and statistical methods to describe the distribution of health and disease in populations and the determinants of that distribution.

Molecular epidemiology combines the methodologies of molecular biology, microbiology, and other laboratory sciences with population approaches used by epidemiologists and the epidemiologic value of making a difference. Basic science research generally has an outcome of understanding the underlying mechanisms leading to a specific function. By contrast, epidemiology is very pragmatic; epidemiologists identify problems and try to fix them, and then check if the fix worked. This pragmatism goes hand and glove with empiricism: if something works, why it works is of less interest than applying it to fix the problem at hand. While this pragmatic approach can be wildly successful, there can be unintended consequences from an empirical approach. Unintended consequences frequently arise from the indirect effects of an intervention. Understanding the underlying mechanisms helps identify indirect effects. When we understand both direct and indirect effects of an intervention, we can more accurately predict when and how to apply it. Thus the merger of molecular biology with epidemiology is potentially even more powerful than the simple combination of laboratory tools with epidemiologic approaches might suggest, because molecular tools enable the epidemiologist to explore the underlying mechanisms leading to a problem of interest, and to use that understanding to better address the problem of interest.

Although the potential is great, fully integrating molecular biology with epidemiology is not easy. Interdisciplinary projects are challenging. To be successful, collaborators must learn each other's jargon and respect the strengths and be cognizant of the weaknesses of each other's disciplines. There can be arrogance on both sides; someone who has never worked in a laboratory may not appreciate the tremendous time, effort, and scientific acumen required to test out a new piece of equipment. Similarly, someone who has never conducted an epidemiologic study, nor managed, integrated, and analyzed vast amounts of data measured using different instruments that vary in quality may believe it is all common sense. On the flip side, an epidemiologist may assume that what comes from the laboratory is correct, and a laboratorian might think the same of something that comes from the computer.

## THE PURPOSE OF THIS BOOK

The purpose of this book is to explore the synergies that emerge from using molecular tools in epidemiologic studies and epidemiologic approaches in molecular studies, and the challenges of conducting a study that integrates the two. My intention is to give the reader an understanding of the challenges of designing, conducting, and analyzing molecular epidemiologic studies. The book covers enough molecular biology for an epidemiologist to read the literature and enough epidemiology to do the same for a microbiologist or molecular biologist. The substance of the text is on how to marry molecular biology with epidemiology.

Molecular biology is currently a rapidly evolving field; technological development is continuing at a ferocious pace. Any text that reviews these technologies will be out of date by the time publication is achieved. However, most of these technologies represent iterative rather than paradigm shifting changes. They are better, faster, or more efficient ways to do what can already be done. These tools make it possible to consider testing large number of individuals – which is required for epidemiology. It also makes it possible to limit the exposition in this book to the core techniques used in molecular biology and focus on the more difficult discussions that must occur for molecular epidemiology to succeed, regardless of what technologies are available. That is, how to identify the correct technique to address your research question, how the available technology frames what research questions might be asked, and how an epidemiologic study is conducted.

Though molecular epidemiology is often envisioned as measuring biological parameters in population studies using molecular tools, population approaches are increasingly incorporated into microbiology and molecular biology. Modern molecular tools have revealed that microbes are populations, and the individuals within those populations vary in ways that influence the ability to be transmitted, persist, acquire genetic changes, and cause disease. To address these questions requires epidemiology.

## WHO THIS BOOK IS FOR

This book was developed for teaching the integration of epidemiology with molecular biology to senior undergraduates or masters' students. I assume students have some knowledge of basic biology, and have been introduced to thinking in terms of populations.

## HOW TO USE THIS BOOK

To truly learn molecular epidemiology, it is optimal to have a training program that includes working in the laboratory, as well as on the design, conduct, and analysis of an epidemiologic study that integrates molecular tools. These experiences cannot be encompassed within the covers of a textbook. What a textbook can do is to provide a context for understanding and interpreting what is read in the scientific literature and a foundation for subsequent practical experience in the laboratory and in the field. The first three chapters give the context, summarizing the history of incorporating laboratory methods in epidemiologic studies, and presenting examples of how molecular tools are applied in epidemiology. As I assume that students using the book may come from a variety of backgrounds, Chapter 4 is a primer on epidemiology and Chapter 5 a primer on molecular biology. These chapters can be safely skipped by students that have already covered them elsewhere. Chapters 6 and 7 present molecular tools and how to choose an appropriate tool for the research question. Chapters 8 through 10 discuss how integrating molecular tools in epidemiologic studies affects the design and conduct of epidemiologic studies. Chapter 11 presents some general analytical strategies. The focus is on the challenges of integrating data across scales, from the molecular to the population. Chapter 12 considers the ethical concerns that arise in molecular epidemiologic studies. In the final chapter, some future opportunities are discussed. Although the chapters build on each other, the instructor may find reading chapters in a different order better suits their target audience. The chapters are sufficiently stand-alone that this is practical.

Depending on the students, the instructor may wish to supplement the text with a variety of other experiences. Available on the World Wide Web are numerous videos that show laboratory techniques. I ask students to find them and present them in class; this is always an enjoyable exercise. For excellent educational resources on genetics, genomics, and other omics, I suggest the learn genetics website (http://learn.genetics.utah.edu/), genomics education website (http://www.genomicseducation.ca/), and the science primer on the National Center for Biotechnology Information website (http://www.ncbi.nlm.nih.gov/About/primer/). For my own course directed at masters' students, the course revolves around designing a molecular epidemiologic study. In a series of homeworks, students explore the literature on the epidemiology of a condition, the molecular tools available to measure the outcome or exposures, and the associated ethical issues. The final paper is a grant proposal.

Writing a textbook is an act of hubris. My decision to attempt it was based on my frustration in finding appropriate materials to teach a course on the molecular epidemiology of infectious diseases that addressed the issues covered here, the interface between molecular biology and epidemiology, rather than a discussion of techniques or applications. The text is based on lectures developed and discussions arising from teaching molecular epidemiology to masters' students in the Hospital and Molecular Epidemiology program in the Department of Epidemiology at the University of Michigan. I first developed this course in 1997 and have taught it almost every year since then. Therefore, I would like to thank the many students who took my course and helped me clarify my thinking about the field. I also would like to thank Carl Marrs, Associate Professor of Epidemiology at Michigan and my laboratory mentor and collaborator over the past 20 years, who made my transformation into a molecular epidemiologist possible. Lixin Zhang, an Assistant Research Professor of Epidemiology, was the first student Carl and I trained jointly in molecular epidemiology and now is a wonderful collaborator; what I have learned from Dr. Zhang about laboratory techniques during his training and since has far exceeded the reverse. I have had many other outstanding students working with me over the years, too numerous to list here, who have also helped me solidify my thinking. Thank you. I would also like to acknowledge my former Department Chair, Hunein F. Maassab, inventor of the FluMist vaccine, who kindly gave an assistant professor the physical and intellectual space required to become a molecular epidemiologist.

Each chapter of this text was reviewed by one or more anonymous reviewers. Their comments were invaluable and I truly appreciate their time and effort. I also very much appreciate the editorial staff at Elsevier who arranged for reviews and editorial assistance. The book is much better for it. Thanks also to my research staff, Usha Srinivasan and Dawn Reed, who put up with me during the writing of this text, and Anna Weaverdyck for providing the outstanding secretarial support that brought the text to closure.

Writing a book takes emotional as well as intellectual support. I am extremely grateful to my husband, Mike Boehnke, for his love and unwavering faith in my abilities to bring this book to completion.

# CHAPTER 1

# Introduction and Historical Perspective

## 1.1 INTRODUCTION TO MOLECULAR EPIDEMIOLOGY

Molecular epidemiology is the discipline that combines molecular biology with epidemiology; this means not merely using molecular techniques in epidemiology or population approaches in molecular biology, but a marriage of the two disciplines so that molecular techniques are taken into account during study design, conduct, and analysis. There is no consensus definition for the term *molecular epidemiology* in the literature (see Foxman and Riley[1] for a review). The term molecular epidemiology emerged apparently independently during the 1970s to early 1980s in the literature of three separate substantive areas of epidemiology: cancer epidemiology, environmental epidemiology, and infectious disease epidemiology. Although these separate substantive areas agree that epidemiology refers to the distribution of disease in a population and the determinants of that distribution, the different literatures present conflicting definitions of what makes a study a "molecular" epidemiologic study. In cancer and environmental epidemiology, molecular is defined almost exclusively in terms of biomarkers. However, biomarkers are only one type of molecular measure, and this definition ignores the many applications of molecular methods in genetic and infectious disease epidemiology. In the microbiology literature, molecular epidemiology has become synonymous with the use of molecular fingerprints – regardless of whether the study was population based or met other criteria consistent with an epidemiologic study. Moreover, the molecular tools available, and the potential for applications for studies of populations, have changed substantially since the term molecular epidemiology was coined.

Since the 1980s there has been an explosion of molecular techniques and of technologies that enable their application to large numbers of individuals – a requirement for epidemiology. In the 1980s, the identification of a single bacterial gene would warrant a dissertation. Now we can obtain the entire genetic sequence of a bacterium, such as *Escherichia coli* whose genome is ~5.5 million base pairs, in a few days (although making sense of the sequence takes a good deal longer). There are databases of genetic sequence of humans, mice, and other animals, and of major human pathogens whose content is growing daily. These databases enable sequence comparisons within and among species, giving insight into possible functions of new genes and the relationships among species. We have vastly improved techniques for determining when genes are turned on and off and under what circumstances, making it significantly easier to characterize proteins. Further, we can measure how the environment changes genome function; these changes, termed the *epigenome*, can be inherited. These advances all deal with material on the molecular level, and "molecular" has become a synonym for modern molecular techniques that characterize nucleic and amino acids, sometimes including metabolites (the "omics": genomics,

Molecular Tools and Infectious Disease Epidemiology.
© 2012 Elsevier Inc. All rights reserved.

transcriptomics, proteomics, metabolomics). Though apparently very broad, this definition of "molecular" excludes many laboratory techniques applied to biological material that might be usefully included in the study of the distribution and determinants of population health and disease (the definition of epidemiology). Thus, for the purposes of this text, molecular is defined as any laboratory technique applied to biological material. However molecular is defined, for a study to be molecular epidemiology, laboratory techniques must be integrated with epidemiologic methods; this integration has profound implications for the design, conduct, and analysis.

Molecular biological techniques enhance measures of diagnosis, prognosis, and exposure, reducing misclassification and increasing power of epidemiologic studies to understand the etiology. However, molecular measures may impose strict requirements on data collection and processing, so that the choice of measure dictates the epidemiologic study design. If the measure of the construct of interest is time dependent or storage sensitive the design is constrained to collection at the relevant time point and the conduct must enable rapid testing. Molecular techniques are generally highly sensitive, specific, and discriminatory; this can increase the power of a study reducing the sample size required. The resulting measures may also require different types of analysis. The investigator must understand what the laboratory result is *really* measuring: it may detect acute exposure or exposure any time in the past. The analysis and interpretation must be adjusted accordingly, as associations with acute exposure predict acquisition, but any time in the past also reflects survival. Thus, a molecular epidemiologic study differs from an epidemiologic study that uses molecular techniques: a molecular epidemiologic study represents a true merger of the disciplines. This implies that the application of epidemiologic methods to the laboratory is also molecular epidemiology. When epidemiologic methods are applied in a laboratory setting, the focus on representative samples and population distributions illuminates the heterogeneity of microbial populations, and of human immune response to those populations, leading to more nuanced interpretations of results from model organisms.

Molecular techniques make it possible to distinguish between infectious agents of the same species with discriminatory power that is far beyond that possible using phenotypic comparisons. This ability has enabled more definitive identification of sources of disseminated food-borne outbreaks (*E. coli* 0157:H7 spread by spinach), demonstration of criminal intent (intentional infection with HIV), and lead us to rethink our understanding of the epidemiology of infectious agents (transmission of *Mycobacterium tuberculosis*). By characterizing the genetics of infectious agents, we have gained insight into their heterogeneity, and rapidity of evolution, highlighting why some previous vaccine development efforts have been unsuccessful. The ability to measure host response to infectious agents has also revealed that our theories about the extent and duration of immunity is somewhat different from that previously thought; lifetime immunity for some infections may only result from boosting from subclinical infection; as we bring infectious agents under control, vaccination schedules must change commensurately. Finally, we are increasingly able to identify the role of infectious agents in the initiation and promotion of previously classified chronic diseases.

When successful, molecular epidemiologic studies help to identify novel methods of disease prevention and control, markers of disease diagnosis and prognosis, and fertile research areas for identifying potential new therapeutics, vaccines, or both. While the integration of the molecular with epidemiological can be very simple, for example, using a laboratory diagnostic measure rather than self-report in an epidemiologic study or describing the distribution of a genetic variant in a collection of bacterial strains, the ultimate success of molecular epidemiologic studies depends upon how well the concerns of each field are integrated. Thus, incorporating a molecular tool that measures the desired outcome or exposure is not sufficient; the strengths and limitations of the chosen measure must be considered in the design, conduct, analysis, and interpretation of the study results.

Within an infectious disease context, molecular epidemiology often refers to strain typing or molecular fingerprinting of an infectious agent; within microbiology, molecular epidemiology generally refers to phylogenetic studies. The field of seroepidemiology, screening blood for past exposure to infection, also falls under the umbrella of molecular epidemiology. However, the realm of molecular epidemiology is much larger, and the potential is much broader than strain typing or phylogeny or testing sera for antibodies, because infectious disease includes two each of genomes, epigenomes, transcriptomes, proteomes, and metabolomes, reflecting the interaction of the infectious agent with the human host. Molecular tools now make it possible to explore this interaction.

Molecular tools are increasingly integrated into epidemiologic studies of environmental exposures, cancer, heart disease, and other chronic diseases. Thus, although the examples in this book all relate to infectious diseases, many of the underlying principles hold for molecular epidemiologic studies of noninfectious diseases. In the remainder of this chapter, I give an historical perspective on the use of molecular tools in epidemiology, then some examples of the range of molecular epidemiologic studies focusing on infectious disease. I close with a discussion of what distinguishes new molecular tools from those used previously, and what distinguishes modern molecular epidemiology from previous studies using laboratory methods.

## 1.2 HISTORICAL PERSPECTIVES

Both in historic and contemporary studies, the inclusion of laboratory evidence strengthens the inferences made using epidemiologic techniques. One epidemiologic hero is John Snow, who identified a strong epidemiologic association between sewage-contaminated water and cholera. Despite extremely well-documented evidence supporting his arguments, his findings remained in doubt for some time. Max Von Pettenkofer, 1818–1901, a contemporary of Snow and also an early epidemiologist, is related, in a perhaps apocryphal story, to have remained cholera free despite drinking a glass containing the watery stool of someone with cholera. From a modern perspective, this demonstrates the importance of infectious dose and host immunity on disease pathogenesis. Snow's conclusions of a causal link were not generally accepted until 25 years after his death, when the cholera vibrio was discovered by Joseph Koch. This discovery enabled Koch to definitively demonstrate the causal relationship between the vibrio and cholera.[2] The strategy of isolating an organism from an ill individual, showing it can cause disease in a disease naïve individual, and then reisolating it as described in the landmark postulates of Henle and Koch, reflects how incorporating laboratory methods enhances our ability to make causal inferences about disease transmission and pathogenesis from even the most carefully researched epidemiologic evidence. Although our understanding of microbiota as a complex ecologic system increasingly undermines the value of the Henle–Koch postulates, the postulates were critical for establishing the causal role of microbes in human health.

Early epidemiologists made tremendous strides with what are now relatively simple molecular tools, such as using microscopy for identification, showing agents not visible by microscope cause disease ("filterable viruses"), and detecting protective antibodies with hemagglutination assays. For example, Charles Louis Alphonse Laveran identified the protozoan that causes malaria using microscopy.[3] Charles Nicolle demonstrated that lice transmitted typhus by injecting a monkey with small amounts of infected louse. Nicolle also observed that some animals carry infection asymptomatically.[4] Wade Hampton Frost used the presence of protective antibodies in the serum of polio patients to monitor the emergence of polio epidemics.[5]

Modern molecular biological techniques, such as those used in genomics, make it possible to distinguish between infectious agents of the same species with much finer discrimination than is possible using phenotypic comparisons. Increases in discriminatory power enable more definitive identification of reservoirs of infection and linkage of transmission events, such as identification of the source of a widely disseminated food-borne outbreak. Characterizing the genetics of human pathogens has revealed the tremendous heterogeneity of various infectious agents and the rapidity with which they evolve. This heterogeneity and rapid evolution helps explain our difficulties in creating successful vaccines for the more heterogeneous organisms, such as *Neisseria gonorrheae*. Molecular analysis has also revealed the role of infectious agents in the initiation and promotion of previously classified chronic diseases, such as cervical cancer. Further, molecular tools have enhanced our understanding of the epidemiology of infectious diseases by describing the transmission systems, identifying novel transmission modes and reservoirs, identifying characteristics of the infectious agent that lead to transmission and pathogenesis, revealing potential targets for vaccines and therapeutics, and recognizing new infectious agents. Molecular tools also make it possible to characterize microbial *communities* (also called microbiota) found in the environment and in and on the human host, and to describe how they influence health and disease. Instead of reducing the disease process to the pathogen that ultimately causes disease, by characterizing the microbiota we can examine how the presence of other microbes may moderate the ability of a pathogen to be transmitted, express virulence factors, or interact with the human host response and thus cause disease. Considering microbes as communities requires drawing on ecological theory, but it helps epidemiologists understand why sewage contamination of drinking water usually leads to an outbreak by a single (rather than multiple) pathogen, how infection by a virus might enhance bacterial infection, or why no single pathogen has been linked to some diseases, like inflammatory bowel disease, even though the condition is characterized by an apparent disruption in normal microbiota. The combination of molecular tools with epidemiologic methods thus opens new opportunities for understanding disease transmission and etiology, and provides essential information to guide clinical treatment and to design and implement programs to prevent and control infectious diseases.

## 1.3 LANDMARK MOLECULAR EPIDEMIOLOGIC STUDIES

The integration of molecular techniques with epidemiologic methods has already solved many mysteries. For example, the epidemiology of cervical cancer is very analogous to that of a sexually transmitted infection, but it only was with the development of modern molecular techniques and their application in epidemiologic studies that a sexually transmitted virus, human papillomavirus, was identified as the cause of cervical cancer.[6,7] During the early part of the epidemic of HIV, it was observed that many individuals with AIDS had Kaposi sarcoma, a rare type of cancer. The epidemiology and natural history of the sarcoma was different from cases of Kaposi sarcoma found in individuals without HIV. The epidemiology suggested that Kaposi was caused by an infectious agent, but no agent could be grown using standard culturing techniques. However, using a case-control design and molecular tools – which did not require growing the infectious agent – a virus was identified, now known as herpes simplex virus 8 or Kaposi sarcoma virus.[8] Similarly, molecular typing confirmed epidemiologic observations that tuberculosis transmission can occur in relatively casual settings,[9] and differentiated diarrheas caused by *E. coli* into pathotypes, which were distinguishable epidemiologically.[10] The powerful combination of molecular tools with epidemiology led to the rapid discovery of the cause of new diseases, such as severe acute respiratory syndrome (SARS).[11] Combining evolutionary theory, molecular techniques and standard epidemiologic methods confirmed that a dentist most likely deliberately infected some of his patients with HIV.[12] Many of these examples will be examined in more detail throughout the text; here, the power of merging molecular tools with epidemiology is illustrated using posttransfusion hepatitis.

Much of what we know about hepatitis comes from studies of posttransfusion hepatitis. Because transfusion is a well-defined event, an association could be made with exposure to blood and blood products and differences in incubation period observed. Also, the epidemiologic strategies could be used to help identify or confirm the etiology. Hepatitis means inflammation of the liver; the inflammation can be caused by alcohol and drug use in addition to several infectious agents. Many hepatitis symptoms are nonspecific – malaise, muscle and joint aches, fever, nausea or vomiting, diarrhea, and headache. The more specific symptoms pointing to liver inflammation, such as loss of appetite, dark urine, yellowing of eyes and skin, and abdominal discomfort, occur regardless of cause, and thus cannot be used to distinguish among them. The incubation period is variable, as short as 14 days for hepatitis A and as long as 180 days for hepatitis B. The multiple etiologies, range of incubation periods, and often nonspecific symptoms made the etiology and associated epidemiology for each etiology difficult to discern. The initial observations suggesting an infectious etiology occurred in the 1880s when hepatitis was noted to follow blood transfusions and injections (remember that needles were routinely reused until the 1980s). But transmission by blood serum was not definitely demonstrated until outbreaks of hepatitis were noted following vaccination for yellow fever in the 1940s (some of the vaccines used human serum).[13]

Before 1970, human blood and blood products used therapeutically had two sources: paid and volunteer donors. Reasoning that an infectious cause of hepatitis might occur more frequently among paid rather than volunteer blood donors, a retrospective epidemiologic study was conducted in 1964 comparing the risk of posttransfusion hepatitis among those receiving blood from the different donor groups. The difference in incidence was substantial: 2.8/1000 units for paid donors versus 0.6/1000 units for volunteer donors. This observation was followed by a randomized trial in which surgical patients were randomly allocated blood from paid versus volunteer donors. Half (51%) of those receiving commercial blood but none receiving volunteer donor blood developed hepatitis. This is strong epidemiologic evidence of an infectious etiology, identified by epidemiology. However, it was not until hepatitis could be detected that it was confirmed that hepatitis was infectious, that there were multiple transmission modes, and that viruses of very different types could cause hepatitis. Nonetheless, even before the detection of hepatitis viruses, the evidence from these epidemiologic studies was used to change medical practice: by eliminating commercial donors the incidence of posttransfusion hepatitis was decreased by 70% (Figure 1.1).[14,15]

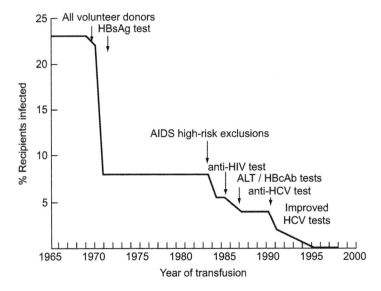

**FIGURE 1.1**

Percent of transfusion recipients infected with hepatitis, by year. *Source: Reproduced, with permission, from Tobler and Busch.*[14]

5

In 1965 the Australian antigen, a marker of hepatitis B, was discovered among individuals with hemophilia who had multiple blood transfusions. Because the antigen caused a reaction in serum from an Australian Aborigine, it was named Australian antigen.[16] After the antigen was associated with what was known as "long incubation period" hepatitis by Alfred M. Prince[17,18] the U.S. Food and Drug Administration mandated screening of blood for hepatitis B in 1972. This further reduced the incidence of posttransfusion hepatitis by 25% (Figure 1.1). It also provided impetus for further studies to identify infectious causes of hepatitis. With the identification of the cause of hepatitis A, a picovirus, it was possible to demonstrate that there were additional causes of posttransfusion hepatitis not attributable to either type A or B. Although it took 10 years and new molecular techniques to identify hepatitis C,[19] the ability to classify types A and B made it possible to learn quite a bit about the epidemiology of non-A, non-B hepatitis, including conducting outbreak investigations.[20] The identification of two surrogate markers of non-A, non-B hepatitis and subsequent inclusion of these markers in screening the blood supply reduced posttransfusion hepatitis rates from a range of 7% to 10% to 2% to 3%, and further reductions followed the implementation of an anti–hepatitis C test.

The history of posttransfusion hepatitis illustrates the great strengths of combining molecular biology with epidemiology. Epidemiologic observations led to the hypothesis that hepatitis could have an infectious cause, and that this infectious cause might occur with different frequency among paid and volunteer blood donors. Limiting the blood supply to volunteer donors dramatically reduced the incidence of posttransfusion hepatitis in the absence of any testing. However, until molecular tests were available, the types of hepatitis could not be distinguished, and the number of infectious agents causing hepatitis was not known. Laboratorians enhanced their searches for additional causes by selecting for testing specimens from individuals with hepatitis of unknown etiology. Finally, the availability of molecular tests enabled screening of the blood supply, enhancing the public's health.

## 1.4 WHAT MAKES MODERN MOLECULAR TOOLS DIFFERENT?

Modern molecular tools can detect trace amounts of material from small amounts of sample at a speed and cost unimaginable just a decade ago. It is possible to use the polymerase chain reaction to amplify nucleic acid from a drop of blood dried on filter paper and identify the genetic sequence of the animal from which the blood was taken. Newer technologies, such as pyrosequencing, enable the rapid sequencing of genetic material, so the entire sequence of a bacterial strain can be determined in less than a week. Modern techniques, such as immunoassays, can be exquisitely sensitive, enabling the detection of material present at parts per trillion. High throughput techniques, such as microarrays, enable rapid testing of large numbers of samples (1000 plus) or the testing of one sample for large numbers of markers. Consider: the time from the first isolation of the coronavirus that causes SARS in 2003 to the determination of the entire genetic sequence of ~29,700 pairs was completed less than 8 weeks following the isolation of the viral RNA.[21]

Molecular tools continue to be developed at a rapid pace, and are increasingly available as commercial kits. This makes it possible to not only characterize populations in a new way, but also revisit and increase our understanding of well-described phenomena, providing a tremendous opportunity for the epidemiologist. However, these kits cannot be used blindly; their reliability and validity must be assessed in the population of interest, and the investigator should understand the variability of the test and sensitivity to variation in time of collection, storage, and processing and design the study accordingly. To do so requires both epidemiologic and laboratory expertise.

## 1.5 HOW MODERN MOLECULAR EPIDEMIOLOGY DIFFERS FROM TRADITIONAL EPIDEMIOLOGIC STUDIES USING LABORATORY METHODS

Laboratory methods are often included in epidemiologic studies. In an outbreak investigation, laboratory methods enhance case definition by confirming disease diagnosis. If the same infectious agent can be isolated from the epidemiologically identified source, the investigation is confirmed. Seroepidemiology studies use laboratory measures to describe the prevalence of previous exposure to an infectious agent. The purpose of a clinical epidemiologic study might be to identify early markers of disease or predictors of prognosis that can be identified in the laboratory. In environmental epidemiologic studies, laboratory assays may be used to measure exposure. In the ideal case, all these studies would be molecular epidemiology studies, not just because they use molecular tools, but because the choice of molecular tool and the implications of that choice are accounted for in the design, conduct, and analysis. Unfortunately, this is often not the case, which weakens any inferences that might be made from the study.

A typical example is using a standard laboratory test for diagnosis. Is the study participant sick or not? Consider the disease strep throat. Strep throat comes with a sore throat, white or yellow spots on the throat or tonsils, and high fever. This sounds definitive, but the same clinical presentation can include other diseases, including infectious mononucleosis. Everyone who has strep throat, by definition, has group A *Streptococcus* growing in their throat. However, not everyone who has group A *Streptococcus* growing in their throat has a strep throat, even if they have a sore throat (the symptoms might be due to a viral infection). Further, there are several ways to detect the presence of group A *Streptococcus*, which vary in sensitivity, specificity, and type of information provided. Some give a yes/no answer, and others give information on the amount of bacteria present. Others test for specific characteristics of the bacteria, such as presence of a toxin *Streptococcus* can produce. If the individual has been treated recently with an antibiotic, it is likely that no group A *Streptococcus* will be detected by culture, but later, group A *Streptococcus* might be cultured from their throat! So, when and if the individual has symptoms, symptom severity and whether the individual has been treated matter. Using a laboratory test alone is not definitive for diagnosis, but it is necessary. However, if the study is to determine how many individuals are carrying group A *Streptococcus*, it may be sufficient, although the investigator must decide at what level of detection an individual is "carrying" the organism, and if those with symptoms will be considered as a different category. Further, if it matters what type of group A *Streptococcus* is found, another laboratory method may be optimal. This is a relatively simple example, but it illustrates how the research question constrains the molecular tool, and how the molecular tool can constrain the timing of specimen collection. How to choose an appropriate molecular tool, how it can constrain the study design, and what it implies about the conduct and analysis of the results are the focus of this text.

## 1.6 OVERVIEW OF THIS TEXTBOOK

This textbook is organized into three sections. Section 1, covered in Chapters 1 to 3, presents examples of how molecular tools enhance epidemiologic studies, with a special emphasis on infectious diseases. Chapter 4 is a primer of molecular techniques, and Chapter 5, a primer of epidemiologic methods. Section 2, covered in Chapters 6 and 7, presents technical material about standard molecular techniques. This section is not meant to teach the reader how to conduct specific assays, but to familiarize them with the vocabulary of current techniques, their applications, and any caveats about their use. The last section, covered in Chapters 8 to 11, discusses the implications of adding molecular techniques to epidemiologic studies. This includes discussions of study design, conduct, and analysis. Chapter 12 focuses on the ethical issues that arise from using biological samples. In the final chapter, I discuss possible hot areas for future research and development.

# References

1. B. Foxman, L. Riley, Molecular epidemiology: Focus on infection, Am. J. Epidemiol. 153 (12) (2001) 1135–1141.

2. S.H. Kaufmann, U.E. Schaible, 100th anniversary of Robert Koch's Nobel Prize for the discovery of the tubercle bacillus, Trends Microbiol. 13 (10) (2005) 469–475.

3. R. Zetterstrom, Nobel Prizes for discovering the cause of malaria and the means of bringing the disease under control: hopes and disappointments, Acta. Paediatr. 96 (10) (2007) 1546–1550.

4. T.N. Raju, The Nobel chronicles. 1928: Charles Jules Henry Nicolle (1866–1936), Lancet 352 (9142) (1998) 1791.

5. T.M. Daniel, Wade Hampton Frost: Pioneer Epidemiologist, 1880–1938, University of Rochester Press, Rochester, 2004.

6. N. Munoz, F.X. Bosch, S. de Sanjose, et al., The causal link between human papillomavirus and invasive cervical cancer: a population-based case-control study in Colombia and Spain, Int. J. Cancer 52 (5) (1992) 743–749.

7. L.A. Koutsky, K.K. Holmes, C.W. Critchlow, et al., A cohort study of the risk of cervical intraepithelial neoplasia grade 2 or 3 in relation to papillomavirus infection, N. Engl. J. Med. 327 (18) (1992) 1272–1278.

8. Y. Chang, E. Cesarman, M.S. Pessin, et al., Identification of herpesvirus-like DNA sequences in AIDS-associated Kaposi's sarcoma, Science 266 (5192) (1994) 1865–1869.

9. J.E. Golub, W.A. Cronin, O.O. Obasanjo, et al., Transmission of *Mycobacterium tuberculosis* through casual contact with an infectious case, Arch. Intern. Med. 161 (18) (2001) 2254–2258.

10. R.M. Robins-Browne, E.L. Hartland, *Escherichia coli* as a cause of diarrhea, J. Gastroenterol. Hepatol. 17 (4) (2002) 467–475.

11. C. Drosten, W. Preiser, S. Gunther, et al., Severe acute respiratory syndrome: Identification of the etiological agent, Trends Mol. Med. 9 (8) (2003) 325–327.

12. C.Y. Ou, C.A. Ciesielski, G. Myers, et al., Molecular epidemiology of HIV transmission in a dental practice, Science 256 (5060) (1992) 1165–1171.

13. L.B. Seeff, Yellow fever vaccine-associated hepatitis epidemic during World War II: Follow-up more than 40 years later, in: W.F. Page, (Ed.), Epidemiology in Military and Veteran Populations: Proceedings of the Second Biennial Conference, March 7, 1990, The National Academies Press, pp. 9–18.

14. L.H. Tobler, M.P. Busch, History of posttransfusion hepatitis, Clin. Chem. 43 (8 Pt 2) (1997) 1487–1493.

15. J.C. Booth, Chronic hepatitis C: the virus, its discovery and the natural history of the disease, J. Viral Hepat. 5 (4) (1998) 213–222.

16. G.L. Gitnick, Australia antigen and the revolution in hepatology, Calif. Med. 116 (4) (1972) 28–34.

17. A.M. Prince, Relation between SH and Australia antigens, N. Engl. J. Med. 280 (11) (1969) 617–618.

18. A.M. Prince, An antigen detected in the blood during the incubation period of serum hepatitis, Proc. Natl. Acad. Sci. USA 60 (3) (1968) 814–821.

19. T. Miyamura, I. Saito, T. Katayama, et al., Detection of antibody against antigen expressed by molecularly cloned hepatitis C virus cDNA: Application to diagnosis and blood screening for posttransfusion hepatitis, Proc. Natl. Acad. Sci. USA 87 (3) (1990) 983–987.

20. G. Gitnick, S. Weiss, L.R. Overby, et al., Non-A, non-B hepatitis: a prospective study of a hemodialysis outbreak with evaluation of a serologic marker in patients and staff, Hepatology 3 (5) (1983) 625–630.

21. National Center for Biotechnology Information, Medicine U. N. L. o. SARS CoV [Internet]. Rockville: 2009. Updated: September 26, 2009; Accessed: December 14, 2009. Available at: http://www.ncbi.nlm.nih.gov/genomes/SARS/SARS.html.

# CHAPTER 2

# How Molecular Tools Enhance Epidemiologic Studies

Molecular tools enhance measurement and classification at a level unimaginable even 10 years ago. We can detect trace amounts of materials, where trace is in parts per trillion. We can amplify trace amounts of DNA and determine what organism it came from. We can determine which genes are turned on or off when exposed to a toxin or infectious agent. We can detect signals within, between, and among cells, and how these vary under different conditions. We can measure interactions among the many microbes inhabiting our bodies, and the interactions among these microbes and human cells. Integrating our newfound ability to detect and classify incredibly small amounts with exquisite accuracy is enhancing epidemiologic studies and advancing the science of epidemiology in a wide range of substantive areas.

9

Epidemiology describes the distribution of health and disease in a population and the determinants of that distribution. Thus epidemiologic studies classify individuals into those with and without disease, and those with and without potential determinants of disease in order to identify disease determinants, and gain insight into disease etiology, transmission, and origins. Molecular tools increase the accuracy of these classifications. In addition, molecular tools can identify past exposure, quantitate extent of current exposure, and determine disease stage, and hence more accurately predict prognosis.

Molecular tools also enable epidemiologic studies to move beyond the detection of risk factors and probable transmission mode, to identify mechanisms of disease pathogenesis and describe transmission systems. Well-designed molecular epidemiologic studies have identified new infectious agents and transmission modes, described how often pathogenic mechanisms identified in the laboratory occur in human populations, and linked transmissibility with disease stage.

Every measurement is subject to error from a variety of sources (see Chapter 8). Our goal is to reduce sources of error so the true biological variation can be observed. Molecular tools are, in general, more discriminatory, precise, sensitive, and specific than other measures. This means that molecular tools identify more categories of a measure, and the variation in the measure is small. Also, given that the test is truly positive, the probability of testing positive is high, and given the test is truly negative, the probability of testing negative is high. These characteristics reduce misclassification, increase the power of a study to detect an association if it exists, and enable the investigator to test a hypothesis using a smaller sample size.

This chapter focuses on how improved measurement and classification enhances epidemiologic studies. The chapter begins with a review of misclassification bias. It then discusses, with examples, how reducing misclassification via molecular tools enhances

*Molecular Tools and Infectious Disease Epidemiology.*
© 2012 Elsevier Inc. All rights reserved.

epidemiologic studies, and how enhancing classification moves epidemiology from detection of risk factors to illuminating disease transmission and pathogenesis.

## 2.1 WHAT IS MISCLASSIFICATION BIAS?

In statistics, a bias is the difference between the true value, say of the mean, and what is estimated from the data. A bias might occur because minor variations in measurement result in a slightly different value. Imagine you wish to measure your height. You might stand against a wall, and put a book on your head, and then mark the height of where the book touched your head. Next, using a tape measure you measure the distance to the floor. Depending on how careful you are, if you repeat the measurement procedure several times, your measured height might vary a half inch (or more) in either direction. This is random error; the direction of error is not always greater or smaller. Suppose, however, your measuring tape only allows measurement to the nearest quarter inch, and you wish to be taller. In that case, you might always round the value to the next highest value, rather than randomly up or down. This is a systematic error, because it is systematically higher, not randomly higher or lower. Now consider measuring a group of individuals that you want to divide into two groups: short and tall. At the boundaries between the groups, measurement error can make a difference in assignment. If less than 5.5 feet is short and over 5.5 feet is tall, and measurement can vary by a half inch either direction, some of the time short people will fall in the tall group and vice versa. The more precise (less variation) and accurate a measurement is, the better the assignment to groups.

The bias associated with incorrect assignment to groups is called misclassification bias. The misclassification may occur at random or be due to a systematic error, such as might occur if using a miscalibrated instrument. Unless a measure is perfect, some misclassification will occur. The variance of a measure quantifies the extent of misclassification. This is usually summarized by the standard deviation or standard error of the mean or with confidence intervals. Say we wish to estimate the proportion of newly pregnant women who are susceptible to rubella. In a 2005 study, rubella immunity was tested using an enzyme-linked immunosorbent assay (ELISA) for viral immunoglobulin G (IgG) titers to rubella[1]; 9.4% (95% confidence interval [CI]: 7.57%, 11.23%) of the 973 pregnant women screened were susceptible. The CI tells us that if the study were repeated 100 times, the true value would fall between 7.57% and 11.23%, in 95 out of 100 replicates. "True value" here refers to the value obtained from the test used; the CI reflects the variance of the measure. The extent to which this measure determines that a woman was truly immune depends on how well it measures the true construct of interest.

Using a measure with increased discriminatory power and lower variance decreases misclassification because individuals are more accurately assigned to the correct groups. Discriminatory power refers to the number of groups: a ruler that measures to the nearest sixteenth inch is more discriminatory than a ruler that measures to the nearest quarter inch. The effects of incorrect assignment can be large, totally obscuring a real association. The history of detecting human papillomavirus (HPV), which causes cervical and other cancers as well as warts, is illustrative of how improvements in a measuring instrument can improve our understanding of the epidemiology. HPV is very hard to grow in the laboratory; although antibodies can be detected in blood, serology studies cannot distinguish between current and past infection. It was only once methods based on detecting viral DNA were available that it was possible to begin to understand the epidemiology of HPV and its association with various clinical conditions. Early HPV detection methods were not as good as later methods; as the detection methods improved, the estimated prevalence of HPV rose, and it was easier to detect associations with other factors, because fewer true positives were included with the negatives and vice versa.

Cervical cancer has an epidemiology similar to a sexually transmitted infection. Like a sexually transmitted infection, cervical cancer is associated with having multiple sex partners

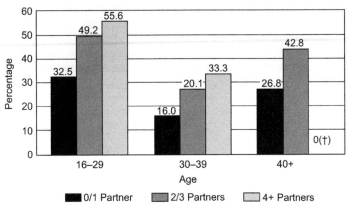

**FIGURE 2.1**

Prevalence of human papillomavirus by age and number of recent sex partners among 404 cytologically normal women in the Washington, D.C., area. *Source: Reproduced, with permission, from Hildesheim et al.*[2]

(or a partner with a history of multiple sexual partners); virtually all women with cervical cancer have engaged in sexual activity. This suggested that the causative factor might be a sexually transmitted infection; indeed, herpes simplex virus was once a key suspect.[3] Once HPV became the prime suspect, it was anticipated that HPV infection would be associated with number of sex partners. A study conducted in 1986–1987 estimated HPV prevalence using a Southern blot test among women in Washington, DC, but found no trend in HPV levels with number of sex partners.[4] In a second study conducted a few years later using another sample from the same study,[2] the investigators tested for HPV using the polymerase chain reaction (PCR), which is better able to detect HPV, if present. The second study showed an increase in HPV prevalence with number of sex partners, and with number of recent sex partners. In addition, it showed that prevalence of HPV decreased with age, but in each age group HPV prevalence increased with lifetime number of sex partners (Figure 2.1). Note that we would expect to observe an even stronger association with acquiring HPV (new cases or incident cases) than with existing cases (prevalent cases). Prevalent cases are a mix of new and existing cases and thus these studies had a source of misclassification not due to the measure but to the study design. Nonetheless, this example shows that misclassification can obscure a strong and consistent association.

A hypothetical example might clarify how misclassification can diminish the strength of an observed effect. Table 2.1 shows test results, and (hypothetical) "true" results, for a test that misclassifies 1% of the true values at random, so that of the 91 persons classified as

**TABLE 2.1** Comparison of Results of Enzyme-Linked Immunosorbent Assay (ELISA) for Viral Immunoglobulin G (IgG) Titers to Rubella to Hypothetic True Values. Test Results Misclassify 1% of True Values at Random

| Test Results | "True" Rubella Susceptibility | | Totals Observed |
|---|---|---|---|
| | Susceptible | No | |
| Susceptible | 82 | 9 | 91 |
| Immune | 1 | 881 | 882 |
| True Totals | 83 | 890 | 973 |

**TABLE 2.2** Comparison of Results of Enzyme-Linked Immunosorbent Assay (ELISA) for Viral Immunoglobulin G (IgG) Titers to Rubella to Hypothetic True Values. Test Results Misclassify 1% of True Values at Random, and 2% of True Immunes Are Erroneously Classified as Susceptible

| Test Results with Systemic Error | "True" Rubella Susceptibility | | Totals Observed |
|---|---|---|---|
| | Susceptible | Immune | |
| Susceptible | 82 | 27 | 109 |
| Immune | 1 | 863 | 864 |
| True Totals | 83 | 890 | 973 |

being susceptible to rubella, 1% are immune, and of the 882 classified as immune, 1% are susceptible.

The probability that a truly positive sample tests positive, called the sensitivity of the test, is excellent (82/83, 99%), as is the probability that a truly negative sample tests negative, called the specificity (881/890, also 99%). The true proportion of those susceptible is 83/973 or 8.5%; so the 95% CI around the observed estimate of 9.4% (91/973), 7.6% to 11.3%, contains the true value. The sensitivity of the test might be increased by changing the cut point for antibody titer considered susceptible (which would decrease the specificity), a trade-off we will consider in more detail in Chapter 8. For planning purposes, such as estimating how much vaccine would be required to vaccinate the susceptible pregnant population, the impact of random misclassification of 1% is small – the observed point estimate is off by less than 1%.

Now imagine that the control samples that are used to determine titer values were improperly prepared so that 2% of the true immunes are classified as susceptible in addition to those misclassified due to random variation (Table 2.2). The proportion observed is only marginally affected: 11%. The sensitivity remains excellent (82/83 or 99%) but the specificity decreases slightly (863/890 or 97%).

Suppose, however, that we wish to identify an association between two variables, for example, how much of rubella susceptibility is attributable to prior vaccination in the past 10 years, and how much to a vaccination that didn't take? Assume that measurement of prior vaccination is perfect and 80% of the susceptible population and 90% of the immune population were vaccinated in the past 10 years (Table 2.3).

If susceptibility is measured without error, the odds of susceptibility among those vaccinated compared to those not vaccinated (odds ratio [OR]) is 2.25. When there is 1% random error, 99% of 83 true susceptibles are classified as susceptible; and 1% of true immunes are classified as susceptible. Thus in the group with random error, the number of susceptible individuals who are vaccinated is (.99*83*.8 + .01*890*.9) or 74, where .8 is the proportion of true susceptibles who were vaccinated and .9 the proportion of true immunes who were vaccinated (we are assuming no error in measuring vaccination), and so forth. The result is that the relative odds of susceptibility among those not vaccinated compared to those vaccinated decreases slightly from 2.25 to 2.10; or it is biased toward the null association of no difference (1.0).

If there is a systematic error putting 2% true immunes among the susceptibles, the bias is in the other direction, putting the OR at 5.32. This particular bias increases the OR because it biases the odds of susceptibility to immune among those not vaccinated upward, from 17:88 (20%) to 35:71 (49%), whereas the odds among those with vaccination remains the same. This type of misclassification is differential, that is, it differs among people with and without

**TABLE 2.3** Demonstration of the Effects of Misclassification on the Association of Immune Status to Rubella with Vaccination Status. Immune Status Measured without Error, with 1% Random Error, and with 1% Random Error and with 2% of True Immunes Are Erroneously Classified as Susceptible. Vaccination Status Is Measured without Error

| Susceptibility without Error | Vaccination Status (Without Error) | | Totals Observed |
| --- | --- | --- | --- |
| | Vaccinated | Not Vaccinated | |
| Susceptible | 66 | 17 | 83 |
| Immune | 801 | 89 | 890 |
| Totals | 867 | 106 | 973 |
| | | OR = 2.25 | |
| **Random Error** | | | |
| Susceptible | 74 | 17 | 91 |
| Immune | 794 | 88 | 882 |
| Totals | 867 | 106 | 973 |
| | | OR = 2.10 | |
| **Random and Systematic Error** | | | |
| Susceptible | 74 | 35 | 109 |
| Immune | 794 | 71 | 864 |
| Totals | 867 | 106 | 973 |
| | | OR = 5.32 | |

a characteristic, depending on who is affected the direction of bias can be away from or toward the null.

## 2.2 HOW REDUCING MISCLASSIFICATION VIA MOLECULAR TOOLS ENHANCES EPIDEMIOLOGIC STUDIES

In the simple examples shown in Section 2.1, we saw how random misclassification can bias an association toward the null, and how differential misclassification can bias results either away from or toward the null. In the examples, only one variable was subject to misclassification; in reality, all variables are measured with some degree of error. Errors compound and the results of the combinations of random error are not predictable. Epidemiologic studies often consider hundreds of variables; even a very simple study will have as many as 10 variables that are included in a regression model as part of the analytic plan. Thus measuring instruments, such as molecular tools that limit random error and reduce misclassification, have the potential to greatly improve inferences made from epidemiologic studies.

Using molecular tools decreases misclassification in general by reducing random error. (Systematic errors remain a problem; see Chapter 8.) The most obvious benefit of reducing random error is increasing our ability to detect an effect if it exists (Table 2.4). An effect may be the difference between two means, the correlation between variables, an odds ratio or

**TABLE 2.4** Positive Impacts of Reducing Misclassification

Increased power to detect an effect if it exists
Enhanced ability to detect smaller effects
Ability to detect same effect with smaller sample size

similar. Consider testing whether the means of two normally distributed populations of the same size are equal using the Student $t$-test.[*] The formula is:

$$t = \frac{\overline{X}_1 - \overline{X}_2}{s_{\overline{X}_1 - \overline{X}_2}} \text{ where } s_{\overline{X}_1 - \overline{X}_2} = \sqrt{\frac{s_1^2 + s_2^2}{n}}$$

where the means of samples one and two are $\overline{X}_1$ and $\overline{X}_2$, respectively, and $s$ is the pooled standard deviation of the two groups. The pooled standard deviation is estimated using the standard deviations for each group and sample size. The larger the value of $t$, the greater the probability that the two means are different. Looking at this formula, we see that the smaller the pooled standard deviation – the measure of variance in the denominator – the larger $t$ will be for any given difference between the two means. We also observe that a larger difference between the means in the numerator will increase the value of $t$, for a given pooled standard deviation. If we look at the formula for the pooled standard deviation, we notice that the sample size, $n$, is in the denominator; increasing the sample size also reduces the variance. Thus, using molecular tools, which are inherently more precise, increases our ability to detect an effect ruling out chance, enables us to detect the same effect with a smaller sample size, and allows us to detect a smaller effect.

To demonstrate these effects, assume we wish to detect if there is a statistically significant difference in mean rubella antibody levels in two populations. Antibody is measured in international units (IU), and a value >15 IU is positive. For statistical tests, a study power (the ability to detect a true effect if it exists) of 80% or greater is considered desirable when the probability of falsely rejecting a true hypothesis (the alpha level or type I error) is 5%. Assume there are two tests available, one with a standard deviation of 10, and another a standard deviation of 15. We test 200 individuals in each population, and the antibody level is 10 in one population and 15 in another. For an alpha level of 0.05, when the standard deviation is 10, our power to detect a statistically significant difference between the groups is 85%; however, when the standard deviation is 15, the power is only 51.6%. With the same sample size, decreasing the standard deviation increases the power. What if the difference between the groups was only 2 IU? In that case, the power to detect the difference for a standard deviation of 10 decreases to 51.6%, but if the standard deviation is 7, the power is 81.5%. Increasing the precision of the measure increases our ability to detect a smaller effect. If the precision of the measure were further reduced to 5, only 99 individuals would be required per group to detect a statistically significant difference of 2. All calculations presented here were done using OpenEpi[5]; power and sample size are presented in more detail in Chapter 11.

## 2.3 WAYS MOLECULAR TOOLS ADVANCE THE SCIENCE OF EPIDEMIOLOGY

Incorporating molecular tools enhances what epidemiologists can do already; reducing misclassification of disease and exposure enhances the ability to detect an association between an exposure and disease. But this is just a beginning. Fully integrating molecular tools into epidemiology opens the door for epidemiologists to advance the science of epidemiology by identifying new indicators of pathogenesis and virulence and new etiological agents that affect humans, and by characterizing interactions among microbes and their hosts and environment that lead to disease (Table 2.5). Some examples of each of these benefits are given in the following sections.

---

[*]The Student $t$-test is described in all standard statistic tests. A $t$-test is used instead of the $z$-test when the sample size is small.

| TABLE 2.5 Ways Molecular Tools Advance the Science of Epidemiology |
|---|
| Direct measures of exposures and outcomes |
| Continuous measures enable detection of dose response |
| Identify disease origins |
| Characterize infectious agents that are difficult to grow |
| Characterize interactions among microbes, their host, and environment |
| Increase understanding of disease pathogenesis and transmission |

## 2.3.1 Direct Measures of Exposures and Outcomes

Epidemiologists are often content to measure a marker of a particular phenomenon or construct rather than directly measure the phenomenon as long as the marker is reasonably predictive. A marker is an indirect measure of something of interest. For example, virtually all epidemiologic studies measure age. Age broadly captures social and biological phenomena; it is generally a marker of one or more factors more directly associated with the outcome of interest. This makes age a very useful variable for generating hypotheses, and for planning health interventions. A diarrheal outbreak limited to children suggests exploration of different venues and foods compared to an outbreak limited to persons over age 65 years or one that affects persons of all ages. If individuals of a particular age group are more susceptible, age might be used to set public health policy (no aspirin to treat fever among persons <21 years) or as part of a guideline for medical diagnosis (gall bladder disease: fair, fat, and 40). But measuring age does not tell us which of the many factors captured is salient nor does it give us much insight into the underlying biological processes.

One advantage of using a molecular tool is that we can more directly measure the construct of interest than is possible using a questionnaire or clinical examination. All other factors being equal, direct measures result in less misclassification than indirect measures, which, as shown in Section 2.2, increases the power of a study to detect an effect. There are several ways to assess that someone is a smoker; it might be observed from the odor clinging to their clothes, we might examine their grocery receipts, the individual might be asked directly, or we might measure a metabolite of nicotine in their blood or urine or saliva. All of these measures have strengths and weaknesses. A smoky odor or grocery receipts are indirect measures. A nonsmoker who lives with a smoker or who goes to venues where smoking is allowed will also smell like smoke. Similarly, a nonsmoker may purchase cigarettes for a member of the household who smokes or for a friend. If asked directly, the individual may lie outright, or underestimate his or her cigarette consumption. Measuring cotinine, a metabolite of nicotine, is also an indirect measure; it tells us that the individual was exposed to nicotine, but not if he or she smoked cigarettes. It also tells us nothing about extended use, although it does give an indication of the extent of exposure. However, it does, within the limit of detection, directly measure nicotine exposure, making it possible to assess the effects of passive smoking on disease risk.

How directly we can measure a construct depends on our understanding of the underlying biological process. An elevated white blood cell count suggests an infectious process without giving much insight into the cause of that process; however, this may be an important indicator in the absence of other evidence. Identifying the type of white blood cell gives clues as to the cause. This, in turn, can lead to additional tests to identify the cause. Before HIV was identified, it was first defined by the clinical presentation: Kaposi sarcoma and other opportunistic infections. Fairly quickly, it was associated with decreased levels of certain types of white blood cells, which became part of the diagnosis. Once HIV was identified, a test for HIV antibodies was developed, which directly measures exposure to the virus. This is similar to the hepatitis example presented in Chapter 1, which described an epidemiologic study demonstrating a different hepatitis rate among individuals receiving blood or blood products from paid compared to volunteer blood donors. That study had no molecular marker and could only infer that there were different types of hepatitis by the differing times between blood transfusion and hepatitis. It was not until there was a molecular marker – the Australian antigen – that we could begin to distinguish between different hepatitis types. Once the Australian antigen was identified, population studies determined the prevalence of the antigen, that it was found in multiple populations, that it could be transmitted in multiple ways, and that the disease presentation was modified by age.

## 2.3.2 Continuous Measures Enable Detection of Dose Response

Many epidemiologic analyses use dichotomous variables, such as disease present or absent, or exposure present or absent. Though analytically convenient, it is likely that the dichotomy results in a loss of information. There is a range of severity for most diseases; similarly, there is a range of exposure. Measuring the ranges enables detection of a dose response: the more exposure, the increasing severity of disease. Measuring ranges also enables detection of a threshold response. Exposure up to a certain level results in little or no disease, with risk increasing dramatically at levels higher than the threshold amount.

Molecular markers of disease or exposure generally are quantitative, enabling detection of dose response and thresholds. Even when the characteristic is inherently dichotomous, a molecular test might reveal new insights. For example, Zhang and colleagues[6] (2001) report a validation of a detection method for dot blot hybridization. The old method read the results by eye, and the other used a machine reader, which gave quantitative results.[6] Dot blot hybridization is a technique where single-stranded DNA labeled with a fluorescent or radioactive probe is used to identify the presence of its complementary sequence. Before the availability of a machine that quantitates the pixels, the results were measured using x-ray film, where a positive appeared as a dark spot. Although human readers had noted that some dots were darker than the others, this effect was noted only semiquantitatively: spots were graded as negative, questionable, positive, and strongly positive. The machine

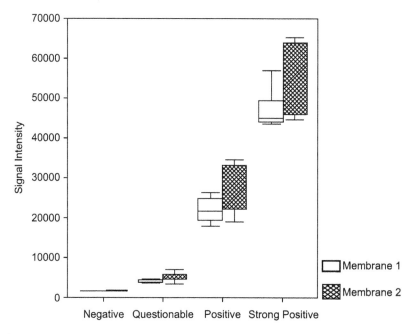

**FIGURE 2.2**

Distributions of replicate values can vary by value. Box plot of DNA hybridization results with a probe that varies in copy number and adherence by allele (*prf* for *E. coli*). Strains were selected from those negative, questionable (later found to be negative), positive, and strong positive. Results from 10 replicates on each strain run in duplicate. The boxes enclose the middle 50% of the data, with a horizontal line at the medial. The "whiskers" are drawn to the maximum and minimum values. *Source: Reproduced, with permission, from Zhang et al.[6]*

enabled quantitation of the extent of positivity, measured by pixel (Figure 2.2). Upon further analysis, the differences in intensity were determined to result from a biological phenomenon as opposed to technical variation. The explanation was that the bacteria that gave the darkest intensity contained multiple copies of the DNA sequence of interest. Thus enhanced measurement has the potential to increase our understanding of what we observe with a specific measurement and to stimulate further studies to explain the findings.

### 2.3.3 Identify Disease Origins

Molecular tools measure a specific construct: metabolite, DNA, RNA, protein. This ability brings epidemiology rapidly into the realm of biology and biological pathways, taking epidemiology into the black box of risk factors to illuminate etiology. It also stimulates epidemiologists to draw upon existing theories, like evolutionary theory, or to develop new theories to explain observed phenomena. These theories can be translated into testable hypotheses. One important area of research focuses on identifying the origin of infectious diseases. Determining the origin of a disease increases our understanding of the disease's spread and pathogenicity, and provides insight into how to prevent the emergence of similar diseases.

How we live influences our contact with microbes, the ability of microbes to be transmitted among us, and the pressures on microbes to adapt. Humans living in environments where there is close contact with animals, where there are insects that transmit disease, or where food or water is contaminated with sewage have different risks of contact with potentially pathogenic microbes than humans who do not live in these conditions. Living close together in large populations makes it possible for a microbe to more easily move from host to host and thus maintain continuous transmission rather than dying out. Medical treatments, such as antibiotics, select for microbes that are resistant to their effects. By examining the distribution of microbial characteristics among microbial populations found in healthy and ill individuals, and comparing characteristics of similar species, it is possible to gain insight into disease origin. These analyses can also identify microbial characteristics associated with microbial transmission or pathogenesis.

In the past 20 years several new infectious diseases have been recognized.[7] Many of these diseases are caused by microbes that also infect animals. By examining the microbial genome and comparing it to other similar organisms, we can identify their evolutionary relationships, that is, the family tree of the organisms. Tuberculosis is a major cause of morbidity and mortality worldwide; an estimated one third of the world's population has tuberculosis. The *Mycobacterium tuberculosis* complex is a group of related bacteria that cause tuberculosis. A related bacteria, *Mycobacterium bovis*, causes tuberculosis in cattle. Before members of the *M. tuberculosis* complex were sequenced, it was hypothesized that *M. tuberculosis* evolved from *M. bovis*. The theory was that when cattle were domesticated, *M. bovis* adapted to live in the human host. A recent analysis of *Mycobacterium* species identified evolutionary relationships that contradict this belief. Using a phylogenetic analysis of the genomes it appears that the human disease adapted to cattle rather than the reverse.[8] (Phylogenetic analysis uses evolutionary theory to infer order of evolution, and is discussed in more detail in Chapter 7.)

*Staphylococcus aureus* is a bacteria that lives closely with humans; it is commonly found in the nose and skin of humans. *S. aureus* are not all the same; they can be separated into different groups using a genetic typing method, called multilocus sequence typing (MLST). Groups are called sequence types or STs. Using phylogenetic analysis, the STs are grouped into clonal complexes or genetic lineages. *S. aureus* is a major cause of hospital-acquired infections, causing a range of diseases from wound infections to pneumonia. Shortly after the introduction of antibiotics, some *S. aureus* in hospitals became resistant to antibiotics.[9] The first *methicillin-resistant Staphylococcus aureus* (MRSA) appeared shortly after methicillin was introduced in 1959. MRSA are now endemic in hospitals, and have recently emerged

17

in the community. Because MRSA are resistant to multiple antibiotics, infection with MRSA can be very difficult to treat. Resistance to methicillin is encoded by the *mecA* gene, which is found as part of a genetic element called *Staphylococcal* chromosomal cassette, (SCC)*mec*. The (SCC)*mec* contains a gene, *mecA*, that codes for antibiotic resistance and the regulation of *mecA*. Whether MRSA found in hospitals are continuously emerging following antibiotic therapy, or are imported from the outside, is an important question for infection control. If MRSA can emerge spontaneously following treatment with antibiotics, it is harder to control than if there are relatively few genetic lineages containing the gene. A genetic analysis of hospital strains suggests that (SCC)*mec* is found primarily in five lineages, but that some of the lineages have different variants of (SCC)*mec*. A study in Portugal tested the hypothesis that circulating strains of MRSA emerged independently in Portugal versus the alternative that MRSA strains were part of the five major lineages, and were imported. The MRSA lineages found circulating in hospitals in Portugal were not related to methicillin sensitive *S. aureus* lineages found in Portuguese hospitals, but were related to lineages circulating worldwide, suggesting that the MRSA was imported.[10]

### 2.3.4 Characterize Infectious Agents That Are Difficult to Grow

Molecular tools make it possible to detect microbes that cannot be grown easily or at all in the laboratory. This has made it possible to describe the epidemiology of these organisms. HPV is an example (described in Section 2.1). Another example is Whipple disease, which is a rare infectious disease that causes diarrhea and abdominal pain, affects other body organs, and is eventually fatal if not treated. The causative agent, *Tropheryma whipplei*, is very difficult to grow, which makes it hard to characterize using phenotypic methods, although it was identified as the probable cause of the disease after it was visualized in 1961 using an electron microscope. Once it became possible to detect and sequence the genome using PCR, how the bacterium was related to other genera and species could be determined, and a differentiation was made between different strains. *T. whipplei* was subsequently detected in saliva and feces of persons without the disease and in environmental sources, although the reservoir remains to be detected.[11]

*Chlamydia trachomatis* has been recognized as a human pathogen for hundreds of years; it was associated with nongonococcal urethritis in the early 1900s, associated with lymphogranuloma venereum in the 1930s, and associated with ophthalmia neonatorum in the 1950s. This obligate intracellular parasite is very difficult to grow and detect; clinically useful diagnostic techniques that made population screening possible were developed in the 1970s. This subsequently led to the identification of a number of syndromes associated with infection.[12] PCR tests enabled detection of C. *trachomatis* in urine rather than from urethral swabs or cervical samples, which have greatly enhanced diagnosis and screening efforts as well as estimates of prevalence and incidence.

### 2.3.5 Characterize Interactions among Microbes, Their Host, and Environment

Since the ability to isolate infectious agents in pure culture and the subsequent articulation of the Koch–Henle postulates, infectious diseases have been conceptualized as caused by a single microbe. Although this conceptualization has been central to our understanding of the epidemiology and prevention of many diseases, it has obscured the important interactions among microbes that also lead to disease. The presence of one microbe might enhance the growth of another: rhinoviruses that cause upper respiratory infections (colds) stimulate the host to increase mucus production, which changes the environment in the nose and throat. These local environmental changes enable some bacteria to move to the middle ear; risk of an ear infection (otitis media) increases following upper respiratory infection.[13] These types of interactions were harder to detect before nonculture techniques; with culture techniques the emphasis is on detecting a specific organism. By contrast, nonculture techniques can be much more general, enabling detection of genes from all organisms present. This makes it possible to move beyond characterizing interacting organisms hypothesized to jointly effect disease risk, such as upper respiratory viruses and otitis media, to characterizing interactions among microbial communities to gain insight into diseases that seem to stem from a disrupted normal microbiota rather than a specific pathogen.

Bacterial vaginosis (BV) is a disease characterized by low levels of lactobacilli and overgrowth of anaerobic bacteria. There is no one organism or group of organisms that is the cause. Typical symptoms are vaginal discharge and a fishy odor. Though the symptoms may be mild, the disease is associated with an increased risk of acquiring other sexually transmitted diseases, including HIV, pelvic inflammatory disease, and preterm birth. A 2008 study used broad-spectrum PCR to describe the vaginal microbial communities among 28 women with BV and 13 women without BV.[14] Women with BV had, on average, three times the number of taxa observed in women without BV. However, there was a great deal of variability in the microbial communities observed within the disease and nondisease groups. Because the disease is polymicrobial, the authors also speculate that there may be interactions among the microbes present that lead to disease, which will be studied further.

Molecular tools also enable measurement of interactions between a specific microbe and its host or environment. Because the lifetime of most microbes is considerably shorter than that of humans, it is possible to observe how a microbe evolves over time. Even for a microbe that grows relatively slowly, such as *M. tuberculosis*, which takes 4 to 6 weeks to grow to detectable levels in culture in the laboratory, it is possible to detect changes within a relatively short time period. A major selection factor for microbial change is medical therapy. Because treatment for *M. tuberculosis* targets replicating cells, and the bacteria grows very slowly, effective therapy requires prolonged treatment, of 6 months or more. Bacteria that have or develop resistance to the treatment will be more likely to survive. However, mutations causing resistance may impose a cost on the bacteria, such as making it more difficult for the mutated *M. tuberculosis* to be transmitted to others. This idea was tested in a study that used molecular techniques to identify different mutations in *M. tuberculosis* associated with resistance to isoniazid, a common treatment, and measured if there were differences in transmission by type of mutation. Some mutations spread more easily than others, and following an analysis that took into account the effects of clinical and patient characteristics, the authors concluded that the characteristics of an *M. tuberculosis* strain were as important as environmental and patient characteristics on tuberculosis spread.[15] This analysis helps explain why there is variation in the frequency of drug resistant *M. tuberculosis* that could not be accounted for by environmental and patient characteristics.

Hepatitis B virus causes active hepatitis, chronic hepatitis, cirrhosis of the liver, and hepatacellular carcinoma (liver cancer). The association of hepatitis B with liver cancer was made possible by applying molecular techniques in a large scale epidemiologic study. In a classic study conducted by R. Palmer Beasley and colleagues,[16] 22,707 Chinese men were followed in Taiwan. The incidence among hepatitis B carriers was 223 times higher than among noncarriers. Aflatoxin also causes liver cancer. This hepatocarcinogen is produced by *Aspergillus flavus*, a fungus that frequently contaminates foodstuffs. Exposure to aflatoxin can be measured in serum and urine using molecular techniques. In four studies that measured both aflatoxin and hepatitis B exposure, estimates of the increased risk of liver cancer following aflatoxin exposure alone ranged from 2 to 17, following infection with hepatitis B alone from 5 to 17, and for both aflatoxin and hepatitis B virus the estimated risks ranged from 59 to 70.[17] With such a large effect when both factors are present it may seem surprising that this synergy was not identified earlier. However, there are many causes of hepatitis and accurate measurement of aflatoxin exposure is impossible using questionnaires alone; the degree of misclassification was sufficient to obscure this strong effect. (Note: The reader may wonder why the first study demonstrated a much larger effect, 223, compared to these other

19

studies. The effect measure is on a log scale, so the differences are not as large as they seem; further, the confidence intervals for all five of the studies overlap.)

### 2.3.6 Increase Understanding of Disease Pathogenesis and Transmission

As discussed earlier, molecular tools enhance the identification of associations between variables by decreasing misclassification. They also enable a refinement of our understanding of the natural history and pathogenesis of disease, as enhanced measurement allows discrimination between disease stages. The results of enhanced measurement can generate new hypotheses about disease pathogenesis, which then can be tested in the laboratory or in the field. One hypothesis regarding disease pathogenesis is molecular mimicry, which is the idea that a pathogen may copy the structures of human cells onto its cell surface, thereby protecting the pathogen from attack by the human immune system. Epidemiologic studies of *Neisseria meningitidis*, a bacterium that causes sepsis and meningitis, found that strains causing invasive disease were more likely to have a surface capsule than colonizing strains. This led to studies of the capsule, the outer membrane of which was discovered to have a lipopolysaccharide that mimics that found on human red blood cells. This mimicry interferes with both innate and adaptive immunity, used by the human body to protect itself from infection.[18] Although it might be possible to design a vaccine that stimulates antibodies to some of these structures, if the human body begins to attack bacteria based on a structure that mimics itself, the body may also attack itself. The result might be to cause autoimmune disease in some individuals. The cause of rheumatic fever is the body attacking itself following infection with *Streptococcus pyogenes*, the bacteria that causes strep throat. Strep throat is very common, but rheumatic fever is rare. Recent work suggests that rheumatic fever is a result of an interaction between bacteria and host; humans with certain genetic mutations are at higher risk.[19]

Infectious diseases circulate through populations, rising and falling in occurrence, even becoming extinct – sometimes temporarily, other times permanently. Identifying the source of an outbreak and tracking its rise and fall is a standard epidemiologic endeavor. Molecular tools enhance this activity by making it possible to discriminate between outbreak and unrelated strains of the same species. Transmission can be more accurately tracked over time and space, and broadly disseminated outbreaks can be identified. Molecular tools can also reveal transmission that does not lead to disease, which can stimulate studies to identify the host characteristics and other factors associated with disease rather than colonization. They can also help identify the reservoir of disease, the animal or environmental source from which a pathogen is reintroduced into the human population. Since the time of John Snow, cholera has been associated with contaminated water. Yet even in places where rates of cholera are high, cholera disappears and reappears. Studies in Bangladesh, where cholera occurs annually, noted that cholera outbreaks followed seasonal blooms of zooplankton. Subsequently, it was discovered that the cholera vibrio lives and multiplies in small planktonic animals called copepods. By filtering the water with a filter sufficient to remove most zooplankton, cholera was reduced by 48% over controls.[20] Thus increased understanding of the disease reservoir for cholera lead to the development of a novel public health intervention.

### References

1. D.M. Haas, C.A. Flowers, C.L. Congdon, Rubella, rubeola, and mumps in pregnant women: susceptibilities and strategies for testing and vaccinating, Obstet. Gynecol. 106 (2) (2005) 295–300.

2. A. Hildesheim, P. Gravitt, M.H. Schiffman, et al., Determinants of genital human papillomavirus infection in low-income women in Washington, D.C., Sex Transm. Dis. 20 (5) (1993) 279–285.

3. B.S. Hulka, Risk factors for cervical cancer, J. Chronic. Dis. 35 (1) (1982) 3–11.

4. M.H. Schiffman, A. Schatzkin, Test reliability is critically important to molecular epidemiology: an example from studies of human papillomavirus infection and cervical neoplasia, Cancer Res. 54 (Suppl. 7) (1994) 1944s–1947s.

5. A.G. Dean, K.M. Sullivan, M. Soe, OpenEpi: Open source epidemiologic statistics for public health, Version 2.3.

6. L. Zhang, B.W. Gillespie, C.F. Marrs, et al., Optimization of a fluorescent-based phosphor imaging dot blot DNA hybridization assay to assess *E. coli* virulence gene profiles, J. Microbiol. Methods 44 (3) (2001) 225–233.

7. National Institute of Allergy and Infectious Diseases. Emerging and re-emerging infectious diseases [Internet]. Updated: September 5, 2008; Accessed: December 14, 2009. Available at: http://www3.niaid.nih.gov/topics/emerging/list.htm.

8. G.P. Schmidt, H. Loeweneck, Frequency of the retroaortic left renal vein in adults (author's transl), Urol. Int. 30 (5) (1975) 332–340.

9. M.C. Enright, The evolution of a resistant pathogen – the case of MRSA, Curr. Opin. Pharmacol. 3 (5) (2003) 474–479.

10. M. Aires de Sousa, T. Conceicao, C. Simas, et al., Comparison of genetic backgrounds of methicillin-resistant and -susceptible *Staphylococcus aureus* isolates from Portuguese hospitals and the community, J. Clin. Microbiol. 43 (10) (2005) 5150–5157.

11. T. Marth, D. Raoult, Whipple's disease, Lancet 361 (9353) (2003) 239–246.

12. J. Schachter, Biology of *Chlamydia trachomatis*, in: K.K. Holmes, P.-A. Mardh, P.F. Saprling (Eds.), Sexually Transmitted Diseases, third ed., The McGraw-Hill Companies, Inc, New York, 1999, pp. 391–405.

13. H.M. Massa, A.W. Cripps, D. Lehmann, Otitis media: Viruses, bacteria, biofilms and vaccines, Med. J. Aust. 191 (9) (2009) S44–S49.

14. B.B. Oakley, T.L. Fiedler, J.M. Marrazzo, et al., Diversity of human vaginal bacterial communities and associations with clinically defined bacterial vaginosis, Appl. Environ. Microbiol. 74 (15) (2008) 4898–4909.

15. S. Gagneux, M.V. Burgos, K. DeRiemer, et al., Impact of bacterial genetics on the transmission of isoniazid-resistant *Mycobacterium tuberculosis*, PLoS Pathog. 2 (6) (2006) e61.

16. R.P. Beasley, L.Y. Hwang, C.C. Lin, et al., Hepatocellular carcinoma and hepatitis B virus. A prospective study of 22,707 men in Taiwan, Lancet 2 (8256) (1981) 1129–1133.

17. M.C. Kew, Synergistic interaction between aflatoxin B1 and hepatitis B virus in hepatocarcinogenesis, Liver Int. 23 (6) (2003) 405–409.

18. H. Lo, C.M. Tang, R.M. Exley, Mechanisms of avoidance of host immunity by *Neisseria meningitidis* and its effect on vaccine development, Lancet Infect. Dis. 9 (7) (2009) 418–427.

19. L. Guilherme, J. Kalil, Rheumatic fever and rheumatic heart disease: Cellular mechanisms leading autoimmune reactivity and disease, J. Clin. Immunol. (2009).

20. R.R. Colwell, A. Huq, M.S. Islam, et al., Reduction of cholera in Bangladeshi villages by simple filtration, Proc. Natl. Acad. Sci. USA 100 (3) (2003) 1051–1055.

# CHAPTER 3

# Applications of Molecular Tools to Infectious Disease Epidemiology

The application of molecular techniques to epidemiology gives epidemiologists the tools to move beyond risk factor epidemiology and gain insight into the overall system of the disease. For infectious diseases, the system includes the transmission system, pathogenesis and virulence of the microbe, and the interaction of the microbe with the human (and other) host(s) and with the microbiota of the host (microbes that normally live on and in the human body). Thus, when dealing with an infectious disease, there are at least two genomes (sets of transcripts, proteins, and metabolites), that of the microbe causing disease and that of the host(s). Molecular tools enhance outbreak investigation and surveillance, facilitate description of the transmission system, increase understanding of the epidemiology, enable detection of previously unknown microbes, and provide insight into pathogen gene function and host–microbe interaction (Table 3.1). This chapter describes, with examples, each of these applications.

## 3.1 OUTBREAK INVESTIGATION

An important step in all outbreak investigations is setting the definition of what constitutes a case (Table 3.2). Molecular tools enhance case definitions, increasing specificity and reducing misclassification, and are now a standard tool in outbreak investigations. Although it is assumed during an outbreak that a single microbe is causing the clinical symptoms, it is possible that a microbe of the same genus and species but different strain is causing disease during the same time period. Molecular typing can distinguish between outbreak and nonoutbreak strains. Laboratory testing also distinguishes between syndromes with a similar clinical presentation. Laboratory screening can minimize misclassification of asymptomatic cases or cases in an early disease stage as nondiseased. Case definitions can be refined by including the molecular type as part of the case definition; this increases the specificity, reduces misclassification of nonoutbreak cases with outbreak cases, and thus increases the potential for identifying the outbreak source. There are a number of different types

**TABLE 3.1** Applications of Molecular Techniques to Infectious Disease Epidemiology

Outbreak investigation
Surveillance
Describe the transmission system
Increase understanding of the epidemiology
Identify previously unknown or uncultivable microbes
Provide insight into pathogen gene function and host–microbe interaction

23

Molecular Tools and Infectious Disease Epidemiology.
© 2012 Elsevier Inc. All rights reserved.

**TABLE 3.2** Core Steps of an Outbreak Investigation

1. Confirm that it meets epidemiologic definition of an outbreak (any one of the following).
   a. There are more cases than expected (surveillance).
   b. Cases are epidemiologically clustered by time, space, or common behaviors.
2. Consider whether there is ongoing transmission (one of the following).
   a. Did regular contact investigations reveal epidemiologic links or similarities among cases?
   b. Did the laboratory identify a genotyping cluster that confirms the epidemiologic links identified by regular contact investigation?
   c. Did the laboratory identify a genotyping or epidemiologic cluster of lab isolates clustered in time and space where there is discordance between the clinical course of the patient and the laboratory results (false-positive culture)?
3. Define an outbreak-related case.
4. Confirm existing number of outbreak-related cases.
5. Investigate existing outbreak-related cases by reviewing:
   a. Medical records (history, physical, clinical chart, and notes);
   b. Laboratory records (serial results of smears, cultures, drug sensitivities, and other testing);
   c. Genotyping results for all culture-positive cases (if not already done, submit isolates for genotyping).
6. Determine the infectious period for each outbreak-related case based on:
   a. Laboratory results, and
   b. Results of screening of named contacts.
7. Determine the sites and facilities frequented and family and social groups exposed by outbreak-related patients during their infectious periods.
   a. Review information from case-patient interviews and contact investigations.
   b. Review information from medical and public health records.
   c. Review information from the facility logs or records.
8. Determine the exposed cohort of persons at each site/facility who may have been present when an outbreak-related case-patient was present during his/her infectious period.
   a. Review information from case-patient interviews and contact investigations.
   b. Review information from medical and public health records.
   c. Review information from the facility logs or records.
9. Determine the duration by number of hours, days, or weeks for the exposed cohort of persons who may have spent time around an infectious outbreak-related patient.
   a. Obtain information from case-patient interviews and contact investigations.
   b. Obtain information from medical and public health records.
   c. Obtain information from the facility logs or records.
10. Prioritize exposed cohorts for screening.
11. Define elements of and action plan for screening, implementation, and follow-up.
12. Identify resources necessary for action plan to be carried out.
13. Create a media plan to respond to possible inquiries.
14. Assign responsibilities and set deadlines.
15. If necessary, expand screening to include low-priority cohorts after screening high-priority cohorts based on evidence of transmission.
16. Evaluate, treat, and follow up additional infected persons associated with this outbreak.
17. Make and implement recommendations to prevent future outbreaks for particular populations or settings involved.
18. Evaluate outbreak response.
19. Determine whether interventions have effectively stopped transmission in this situation.
20. Identify the lessons learned that could improve the public health response to the next outbreak.

Source: Adapted from the Guide to the Application of Genotyping to Tuberculosis Prevention and Control: Appendix B.[2]

of laboratory tests that provide a molecular fingerprint. Most are based on the microbial genotype (see Chapter 5 for a description of different molecular fingerprinting methods), although phenotypic characteristics – such as serotype and antibiotic resistance profile – are also used. One typing method is called pulsed-field gel electrophoresis (PFGE), and in 2010 it is the standard for typing food-borne outbreaks.

The first step in an outbreak investigation is to confirm that an outbreak occurred, that is, there are more cases than expected or a space–time or behavioral cluster (Table 3.2). Space–time clusters can occur by chance alone, and molecular tools make it possible to distinguish between a cluster caused by different strains of the same species and one caused by a single strain. A cluster caused by a single strain likely indicates an outbreak. *Salmonella* is a common cause of diarrhea, transmitted by the fecal–oral route. An outbreak of a particular strain of *Salmonella* was first identified after a case of salmonellosis was reported to the health department in South Carolina. The first case identified a second case and the putative source – turtles. The turtles and cases all carried the same strain, *Salmonella* Paratyphi B var Java; a comparison of the molecular type with strains previously reported to the surveillance system identified additional cases. The case definition was thus "illness with onset from May 1, 2007, through January 31, 2008, in a U.S. resident yielding a *Salmonella* Paratyphi B var Java isolate with the outbreak PFGE pattern."[1]

Molecular typing confirms that there is ongoing transmission, and verifies epidemiologic linkage identified by contact investigation. Examples of epidemiologic linkage are that the individual had the same disease or syndrome during the appropriate time period, and was linked to other cases in some way. In the salmonellosis outbreak, all cases had contact with turtles; the first two cases had swum together with the pet turtles in a swimming pool. Typing can suggest linkage based on a common molecular fingerprint, and can confirm if the epidemiologically identified outbreak source, such as a food item, contains a microbe of the same genotype causing the outbreak, enhancing causal inference. In the outbreak of *Salmonella* associated with turtles, a total of 107 cases were identified that met the case definition; 72% of cases compared to 4% of controls reported turtle exposure in the week before illness.[1] As confirmation of turtles as the source of the outbreak, samples were collected from six turtles (or water from turtle habitats) belonging to cases in four different states, and all the samples were positive for *Salmonella* of the outbreak type and PFGE pattern.

It is often difficult to distinguish between diseases based on clinical presentation alone. This can complicate outbreak investigations, especially if the symptoms are not very specific. There are many viruses that cause flulike symptoms; classification as influenza based on clinical presentation is specific only during an epidemic when the majority of flulike illnesses are caused by influenza. Even during an epidemic, laboratory confirmation is required, as there can be more than one strain of influenza in circulation. In 2008, there were two predominant influenza A strains in circulation: H1N1 and H3N2. Laboratory confirmation is particularly helpful for classifying individuals with mild or atypical symptoms, and determining the specific type.

Outbreak investigations undertaken by the Centers for Disease Control and Prevention (CDC) are routinely reported in the *Morbidity and Mortality Weekly Report*, which is available at the CDC website. One such investigation[3] investigated a prolonged multistate outbreak of *Salmonella enterica* serotype Schwarzengrund infections in humans that was associated with dog food. The outbreak was first detected by local surveillance of PFGE types of *Salmonella*, which identified a cluster of three cases; CDC was notified. At CDC, a comparison with other reports identified isolates from multiple states with the same PFGE type, which stimulated an investigation. The case definition specifically included infection with the outbreak PFGE type; when epidemiologic evidence pointed to dog food as the source of infection, molecular tools confirmed presence of the outbreak strain in unopened bags of dog food and in environmental samples from the implicated manufacturing plant. While not mentioned in the report, continued surveillance of the PFGE type of *Salmonella enterica* serotype Schwarzengrund infections could be used to confirm that the public health intervention – temporary closure of the plant for cleaning and disinfection – was successful in ending the outbreak.

## 3.2 SURVEILLANCE

Public health surveillance is the "ongoing, systematic collection, analysis, and interpretation of data (e.g., regarding agent/hazard, risk factor, exposure, health event) essential to the planning, implementation, and evaluation of public health practice, closely integrated with the timely dissemination of these data to those responsible for prevention and control."[4] Public health surveillance is a cornerstone of public health infrastructure. Surveillance includes collection, analysis, and dissemination components (Figure 3.1). Data collected include incidence, morbidity, mortality, vaccination, clinical, behavioral, and laboratory data. Collected data are analyzed to monitor disease trends, giving a baseline for detection of outbreaks and epidemics, and to evaluate the effectiveness of public health interventions. Reports are disseminated regularly to decision-makers.

Laboratories are key components of many surveillance systems and essential for infectious diseases. Hospital laboratories may be part of both regional and local surveillance networks. Monitoring of infectious disease isolates identifies time–space clusters of infection; molecular typing distinguishes between microbes of the same species, allowing differentiation between clusters of disease occurring by chance and true outbreaks. Spurious clusters do not have a common source, so their investigation wastes time and resources. Molecular typing helps link disease reports from different geographic areas, as we saw in the outbreak of *Salmonella* associated with turtles presented in Section 3.1. True outbreaks and clusters of the same strain can be traced back to a common source and presumably are amenable to public health intervention. Applying molecular tools to surveillance isolates can also identify new strains with increased virulence or changing patterns of resistance.

Hospitals have high endemic rates of bacterial infection, but the infections are often due to a bacterial strain that was colonizing an individual before entering the hospital, for example, *Staphylococcal aureus*. The prevalence of *S. aureus* colonization among the general population is 32% in the nares,[6] but much higher in patients and personnel in hospitals and long-term care facilities. By typing strains causing infection among patients we can distinguish between a strain from the community and one circulating endemically or causing an outbreak within the hospital. The prevention and control strategies are different in each case, and thus it is important to distinguish between them. Screening new hospital patients for the presence of *S. aureus* and then intervening to prevent spread or self-inoculation can reduce introduction of new strains. For strains already circulating in the hospital, screening hospital personnel and retraining personnel on proper hygiene may be in order. In hospitals there are a variety of surfaces that may be contaminated with potential pathogens that might serve as a source of infection. A study conducted in a 1600-bed hospital in Taiwan explored whether computer keyboards and mice might be a source of pathogens.[7] Though one of three major pathogens was cultured from 17% of the 282 computer keyboards or mice, the PFGE types were all different from the PFGE types of clinical specimens obtained from the same wards, suggesting that good hygiene was sufficient to keep these devices from being a source of infection.

Surveillance can occur on multiple levels: in the United States there are surveillance systems within hospitals, cities, counties, states, and the entire country. By monitoring isolates from time–space clusters for the presence of a common molecular type at multiple levels, we can distinguish

**FIGURE 3.1**
A schematic of surveillance.
*Source: Reproduced, with permission, from Trostle.*[5]

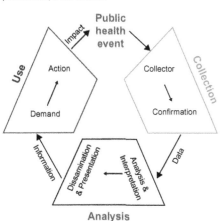

between common-source outbreaks that are local and those that are widely disseminated. Food is frequently distributed widely; molecular tools enable identification of a common cause of disease disseminated through complex food distribution networks that cross large geographic areas. The CDC PulseNet, a molecular subtyping surveillance system for food-borne bacterial disease, monitors *Escherichia coli* O157:H7, *Salmonella, Shigella,* and *Listeria monocytogenes,* and other bacterial pathogens[8] causing disease throughout the United States. In 2006, clusters of a common *E. coli* O157:H7 pulsed-field type were observed at several monitoring sites. An investigation revealed the source of the outbreak as prepackaged, fresh spinach. Once the epidemiologic investigation identified spinach, the public was notified and *E. coli* O157:H7 with the putative pulsed-field type was isolated from an unopened package of spinach from an individual's home. Molecular typing enabled rapid linkage of cases occurring across several states and the identification of the disease source, and it facilitated quick public health intervention – recall of the spinach.[9]

There is also ongoing surveillance for microbes resistant or insensitive to prevailing therapies. Surveillance monitors the emergence and spread of resistance, and provides essential information for effectively treating infection. A cluster of drug-resistant infections is often the first indication of an outbreak, particularly in a hospital setting. Mobile genetic elements that confer resistance can be exchanged between bacteria, even across species, complicating outbreak investigation. Molecular tools can distinguish between a common strain of a single bacterial species or a mobile genetic element conferring antibiotic resistance across strains of the same or even different species. Appropriate intervention should take into account whether a mobile genetic element is being exchanged between species or if there is clonal spread of a single organism. In an outbreak of multidrug-resistant *Pseudomonas aeruginosa* that occurred in a university hospital in Greece, isolates had a novel gene variant coding for resistance. The outbreak was primarily due to clonal spread of the same strain, but a second strain was found to carry the same novel gene variant, suggesting that some of the outbreak was due to the transfer of resistant genes to sensitive strains.[10]

Molecular tools have also been applied to screen biological specimens collected as part of ongoing national databases for the presence of known and newly discovered microbial pathogens. For example, blood samples are collected as part of the National Health and Nutrition Examination Survey, a multistage probability sample of the United States conducted every 10 years. Screening of the samples collected enabled the estimation of the prevalence of various human pathogens, including hepatitis B and C viruses, human herpes virus 8, which causes Kaposi sarcoma, and herpes simplex viruses 1 and 2. These studies provide insight into the frequency of new pathogens, and the distributions of pathogens by spatial–temporal and host characteristics. Such studies are extremely useful for generating hypotheses about transmission systems and potential prevention and control strategies, for evaluating the effectiveness of ongoing prevention and control programs, and observing time trends.

## 3.3 TRANSMISSION SYSTEM

The transmission system of a pathogen determines how pathogens are circulated within a population, and includes the transmission mode, interactions between the pathogen and the host, the natural history of the infection, and interactions between hosts that lead to infection. The emergence and re-emergence of a variety of pathogens highlights the utility of understanding the various transmission systems, because this understanding is central to identifying effective prevention and control strategies. Combining molecular typing methods with questionnaire data can confirm self-reported behaviors, especially important when the validity of self-report may be in doubt, such as might occur during contact tracing for sexually transmitted diseases. Molecular tools also facilitate estimating parameters key to understanding the transmission system, including the incidence, prevalence, transmission probability, duration of carriage, effective dose, and probability of effective contact.

When using simple transmission models to estimate the basic reproductive number ($R_0$), the average number of new cases generated from each infectious case in a fully susceptible population, we need the transmission probability per effective contact, the duration of infectivity, and the rate of effective contact. Molecular tools can usefully be applied in studies estimating each of these parameters. Before the availability of modern molecular tools, our ability to empirically estimate transmission probabilities was limited. Although the transmission of a sexually transmitted infection can be estimated by following couples where one is infected and the other is susceptible, without molecular tools, it is difficult to verify that the transmission event came from within the partnership. For respiratory infections, such as pulmonary tuberculosis, our estimates of the transmission probability and natural history have been based on careful documentation of outbreaks. However, as we have been able to type individual strains, we have determined that tuberculosis cases that previously were considered sporadic and not part of apparent time–space clusters (because the exposure to the index case was very limited) were indeed part of the same outbreak.[11]

Key transmission system parameters are incidence, prevalence, duration of infection, and transmission probabilities (Table 3.3). The accurate estimation of these parameters assumes that we can distinguish between subtypes (strains) of a microbe. As we have increased our ability to type microbes, we have been forced to re-evaluate many of our previous assumptions, because microbes are much more diverse than previously imagined. One prior key assumption was that pathogens are clonal, that is, during active infection all infecting organisms are the same. Genetic studies have emphasized that bacteria exist in populations; there is enough variability within even a clone that the genomic sequence obtained is an average of the population sequence rather than that of a specific individual.[12] Some pathogens, RNA viruses for example, mutate quickly, so that the genetic type of a strain changes over the course of an epidemic, potentially complicating how a case is defined. A second prior assumption was that during an infectious process the pathogenic organism was the one most frequently isolated from the infected site. For many diseases we now know this assumption is false. During a diarrheal episode, the predominant organism isolated from the stool may not be the one causing the symptoms; a toxin-secreting organism occurring at low frequency may be the culprit. For microbes that also are human commensals, such as *Streptococcus agalactiae*, strains causing disease may be different from normal inhabitants, and different strains may have different transmission systems. A third assumption was that individuals are only infected with one strain of a pathogen. This is also false. Individuals can be concurrently infected with different strains of human papillomavirus (HPV), gonorrhea, and tuberculosis.

28

**TABLE 3.3 Key Transmission System Parameters**

| Parameter | Definition | Comments |
|---|---|---|
| Incidence | New cases/individuals at risk/time | Incidence is a rate |
| Prevalence | All cases (new and existing)/all individuals | Prevalence is a proportion |
| Duration of infection | Time from infected until infection is lost | Duration is the reciprocal of the rate that infected individuals lose their infection |
| Transmission probability | The probability (chance) that disease will be transmitted when there is effective contact of a susceptible individual with an infected individual | Can be estimated from the number of transmission events/effective contacts between a susceptible and an infected individual |
| Rate of effective contact | Number of contacts sufficient to transmit disease/total contacts/time | |

These observations have profound impacts on the conduct of future molecular epidemiologic studies of infectious diseases. If the population genetic structure of pathogens is not clonal and the pathogen is not readily isolated, this must be reflected in the sampling of isolates for study. Multiple isolates must be sampled and tested from an individual. For example, if there is a second strain only 5% of the time and the pathogen is uniformly distributed in the sample, we must sample 28 different isolates from an individual to reliably detect the second strain. Further, if a pathogen mutates rapidly within a host, such as HIV, determining the mutation rate will be essential for accurately estimating transmission probabilities and following transmission chains. Moreover, laboratory analyses should take into account the heterogeneity of the organism when selecting isolates for analysis.

Understanding the full extent of the circulation of a particular pathogen is essential for making accurate predictions and determining appropriate prevention and control strategies. The two parameters of greatest interest are incidence, number of persons newly infected during a defined time period, and prevalence, the number of persons infected (Table 3.3). Prevalence includes both new and existing cases and can be measured at a single time point (point prevalence) or over a defined period (period prevalence). Incidence can be estimated by following a cohort and measuring the occurrence of new infections during a defined time period. Strain-specific estimates of prevalence and incidence are essential to our understanding of disease etiology, especially if different strains have different propensities to cause diseases. There are many types of HPV; only a few cause cancer. Before the development of the HPV vaccine, the incidence of infection with HPV type 16, which is a type strongly associated with cervical cancer, was estimated by measuring antibody to HPV 16 in the blood (serology). Women were tested for antibodies to HPV type 16 at the time of their first pregnancy and retested at the time of a second pregnancy. Incident cases were all women who tested negative for HPV type 16 during their first pregnancy that tested positive during their second pregnancy. Seroconversion rates by age (number seroconverted in a specific age group divided by total seronegative at the first pregnancy in that age group) were highest for women younger than 18 years (13.8%); this fell to 2.3% for women age 21 years.[13]

For some diseases, it is possible to distinguish between new and existing cases using molecular tests, so that it is not necessary to follow a cohort over time to estimate incidence. HIV is diagnosed using a test for host response to infection (antibodies), but antibodies do not appear until several weeks after infection. Once antibodies appear, infection can be diagnosed, but the standard test cannot determine how long it has been since infection occurred, so only prevalent cases are detected. Having a test to detect incident cases would be extremely useful for estimating the rate of ongoing HIV transmission with a single survey or as part of a screening program or intervention trial. Molecular tools have been developed to make this possible, and there are several tests available. After HIV status is identified, an additional test is conducted for a substrate that either is present only in those with recent infection or is a marker of extended infection.[14] One such assay uses a branched synthetic peptide (called BED) that enables detection of multiple HIV subtypes; the assay quantitates the proportion of anti-HIV antibody in the serum that increases with time infected. There are still a number of issues with the interpretation of these tests, and it has been noted that biological changes in persons who have been infected for a long time may lead to a high false-positive rate. This is of particular concern in populations with high prevalence. Nonetheless, these tests are an important advance for monitoring HIV and evaluating HIV interventions.[14]

HIV almost invariably leads to detectable disease, but other infections do not. Many common infections, for example, diarrheal diseases, cause disease in only a subset of all infected, or the disease presentation is sufficiently mild that it does not require medical intervention. However, from a public health perspective it would be extremely useful to know the extent of the population infected, because this would enable a more accurate evaluation of the effectiveness of prevention strategies compared to an assessment based on reduction in the number of outbreaks identified. Refinements of molecular tools combined with statistical approaches make it possible to estimate seroincidence. Seroincidence is the number of infections that lead to seroconversion within a defined time interval. The method uses the distribution of known changes in antibody titer since infection to estimate time since last infection, and then converts the time estimates to incidence. This approach has already been used to estimate seroincidence for *Salmonella*, a major cause of diarrheal disease.[15] Though seroincidence cannot be used to estimate disease burden, because not all infections lead to disease, it does provide a useful estimate of the occurrence of new infections in the population.

The duration of infection can be estimated from the prevalence and incidence, presuming that the average duration across strain type is of interest, as prevalence divided by 1 − prevalence = incidence times duration. If the disease is rare, prevalence is approximately incidence times duration. For many pathogens that can cause disease, pathogen presence does not have a one-to-one correspondence with symptoms: the pathogen can be carried asymptomatically. The duration of carriage is important for understanding the transmission system; the duration of disease is useful for estimating disease burden. If duration of carriage is short but incidence is high, an individual might become reinfected with a different strain type, suggesting a longer duration than if strain types were determined. By contrast, if a strain mutates rapidly within the human host, we might underestimate the duration of carriage, an essential parameter for predicting ongoing circulation. Different types of the same species may vary in duration of carriage. *S. agalactiae* is an emerging pathogen that also frequently colonizes the rectum and the vagina. It has nine known serotypes that are also detected by gene sequence (capsular type). The duration of carriage varies by capsular type, and the duration is longer in women than men, presumably because of affinity of the pathogen for the vaginal cavity.[16]

Transmission probabilities are difficult to estimate directly, unless the microbe is transmitted by person-to-person direct contact. For sexually transmitted diseases, the usual design is a couple study where one member of the couple is infected and the other is susceptible. Couples are followed until transmission occurs. Molecular tools are used to verify that both members of the couple were infected with the same strain of the microbe of interest. A couples study design estimated the probability of transmission of herpes simplex virus 1 (HSV1) during pregnancy; among 582 women initially seronegative for HSV1 with an HSV1-positive partner, 3.5% acquired HSV1.[17] If sexual transmission is hypothesized, studying couples at a single point in time, a cross-sectional study, can also be informative. If couples carry the identical molecular type of a microbe more frequently than expected based on the population distribution of the microbe, it is strong evidence for sexual transmission. Sex partners of women with a urinary tract infection are more likely to be co-colonized with the *E. coli* strain that caused the urinary tract infection than a commensal *E. coli* colonizing the woman's rectum,[18] supporting the hypothesis that urinary tract infections can be sexually transmitted. However, dormitory roommates of men and women carrying *S. agalactiae* were no more likely to carry the same strain than expected by chance alone, where chance was estimated based on the distribution of molecular types in the study population.[19]

Molecular tools can also assist in the estimation of contact patterns, by identifying asymptomatic and low level of infections. Contact patterns are a description of the interactions occurring between individuals sufficient to transmit the microbe. For a sexually transmitted disease, contact patterns of a population describe who has sex with whom. Asymptomatic infection is often a key component in maintaining disease transmission. For example, in a study of intrafamily transmission of shigella, asymptomatic carriage increased risk of a symptomatic episode within 10 days by ninefold.[20] Molecular typing can also be used to enrich and validate contact tracing information. The addition of molecular typing to epidemiologic information on gonorrhea cases in Amsterdam identified large clusters of individuals with related strains, individuals infected with different strains at different anatomical sites, and persons with high rates of reinfection.[21]

Applying molecular typing to ongoing or endemic disease transmission increases our understanding of how contact patterns produce observed patterns of disease, revealing novel prevention and control strategies. In addition to characterizing ongoing chains of transmission, molecular typing can clarify who had contact with whom, and who was the source of infection, and thus identify a transmission network. Identifying transmission networks provides essential information for targeting intervention programs, particularly when designing and implementing vaccine programs. Using PCR–restriction fragment length polymorphism typing of the porin and opacity genes of *Neisseria gonorrhoreae* and questionnaire data, a study of successive gonorrhea cases in Amsterstam identified several ongoing transmission chains. The epidemiologic characteristics, including number of sexual partners and choice of same or opposite partners, of patients with different molecular types differed, suggesting that the transmission chains represented different transmission networks.[21] Further analysis revealed that the transmission networks for men who have sex with men, and heterosexuals were essentially separate – a key public health insight for planning interventions. Molecular typing has also improved our understanding of tuberculosis transmission. Until confirmed by molecular typing, tuberculosis was not believed to be transmitted by short-term, casual contact. Several investigations have demonstrated that this assumption is incorrect, because clusters have been associated with use of services at day shelters,[22] and even linked to only a few brief visits to an infected individual's work site.[11] Molecular typing has also demonstrated linkage between apparently sporadic tuberculosis cases, and determined that at least some recurrent tuberculosis is attributable to exogenous reinfection.[23]

## 3.4 INCREASE UNDERSTANDING OF THE EPIDEMIOLOGY OF INFECTIOUS DISEASES

Molecular tools enable us to trace the dissemination of a particular subtype across time and space and thus develop theories of transmission and dissemination; determine the origin of an epidemic and test theories about reservoirs and evolution of a particular pathogen; follow the emergence of new infections as they cross species, testing our hypotheses about the apparent transmissibility and rate of evolution; and follow mobile genetic elements conferring antimicrobial resistance or virulence between strains within a species or between species, and so develop theories about evolution and transmission within the populations of pathogens (Table 3.4).

### 3.4.1 Trace Dissemination of a Particular Subtype Across Time and Space

Microbes that cause human disease are constantly emerging and re-emerging. To prevent and control the spread of infection, we must be able to trace the origin and source of entry of pathogens into the population. By comparing strains we can determine if there have been single or multiple points of entry, and if emerging resistance was from multiple spontaneous mutations or from dissemination of a single clone. Until 2004, only occasional

**TABLE 3.4** Ways Molecular Tools Increase Understanding of the Epidemiology of Infectious Diseases

Trace dissemination of a particular subtype across time and space
Determine the origin of an epidemic
Follow emergence of new infections
Trace mobile genetic elements

isolates of gonorrhea found in Sweden were resistant to azithromycin, and these cases were attributed to acquisition elsewhere.[24] However, in 2004 epidemiologic evidence suggested that domestic transmission might have occurred; this was confirmed by molecular typing. The ongoing transmission of the azithromycin-resistant strain in Sweden has short-term implications for surveillance and long-term implications for treatment recommendations.

*Streptococcus pneumoniae* is a major cause of pneumonia, but also causes meningitis and otitis media. A major human pathogen, it is one of the most common indications for antibiotic use. Resistance to penicillin emerged relatively slowly, but once it emerged it was widely disseminated in relatively few clones as defined by multilocus sequence typing (MLST). By contrast, the recent emergence of *S. pneumoniae* resistant to fluoroquinolones has been due to a diverse set of genetic mutations,[25] suggesting spontaneous emergence following treatment. Because *S. pneumoniae* resistant to fluoroquinolones rapidly followed the introduction of fluoroquinolones, alternative antibiotics will be needed in relatively short order to treat *S. pneumoniae* infections.

### 3.4.2 Determine the Origin of an Epidemic

Molecular tools enable us to trace an outbreak or epidemic back in time to its origin, and back in space to its reservoir. Knowing the origin in time is essential for predicting future spread, and identifying the reservoir for infection is central for controlling disease spread. The use of molecular techniques has solved long-standing mysteries, such as cholera's reservoir between epidemics. The same strains of cholera that infect humans also thrive in aquatic environments, living in zooplankton.[26] During zooplankton blooms the population of cholera vibrio also grows and is more likely to invade the human population.

Molecular tools also provide insight into the origins of infection in highly endemic* populations such as hospitals. The prevalence of methicillin-resistant *Staphylococcus aureus* (MRSA) has been steadily increasing in hospitals in the United States; in 2004 the prevalence among some intensive care units was as high as 68%.[27] However, in the early 2000s, new strains of MRSA emerged among individuals in the community that could not be traced back to hospitals. Genetic typing of the strains confirmed that strains isolated from those who had no epidemiologic linkage with hospitals were genotypically different from hospital strains.[28] More recently, community-acquired MRSA has been introduced into hospitals. Because community-acquired MRSA has, to date, different patterns of antibiotic resistance than hospital-acquired MRSA as well as different virulence factors, there is a clinical benefit in being able to distinguish between the two.[29]

Influenza season comes every year, and we can predict, with reasonable accuracy, which strains will be circulating, enabling preparation and distribution of vaccine. Prediction is based on surveillance of influenza worldwide. The influenza virus mutates as it circulates; the mutations can be modest, known as "drift," such that there is cross immunity with the previous strain; or mutations may be dramatic, where the virus acquires genes by recombining with other strains, known as "shift." Antigenic shifts can occur when human influenza recombines (exchanges genetic material) with other influenza strains. There are influenza viruses that infect humans, fowl, and swine. Pigs have cell receptors that make them susceptible to both avian and human influenza as well as swine influenza, so genetic reassortment between different influenza viruses can occur. It was previously thought that recombination between human and bird influenza within a pig was necessary before an avian flu could infect humans. However, this is not the case. Molecular tools have clarified that avian influenza need not first pass through the pig before jumping to humans, and that strains directly transmitted from birds to humans are often more virulent.[30]

---

*An infection is endemic if there is continued transmission within the population. By contrast, an epidemic is when there are more cases of a specific disease than expected.

In 1918 there was a very severe epidemic of influenza, which, unlike seasonal flu, was most severe in young adults. This virus was different from those seen previously and probably originated in birds. This founding virus, an influenza A H1N1, remains with us (Figure 3.2), and its descendents plague us to this day. The founding virus was introduced to pigs by humans; in 2009 an H1N1 virus was transmitted from pigs to humans.[31] As of yet, we are unable to predict when an antigenic shift will occur or when an avian virus will jump into humans or pigs. However, our ability to trace the flow of specific viral types and their mutations over time provide important information for predicting disease spread and hence for developing effective prevention strategies.

### 3.4.3 Follow the Emergence and Spread of New Infections

SARS was the first new disease to emerge this century. Before its identification, coronaviruses were not considered major pathogens – only 12 coronaviruses were known to infect humans or other animals. The SARS identification led to a search for additional coronavirus pathogens, and ultimately horseshoe bats were identified as the reservoir and civets as the amplification hosts.[32] The time from the initial observation to the sequencing of the virus and development of a diagnostic test was 5 months. The story of the rapid isolation, identification, and sequencing of the coronavirus causing SARS is illustrative of the synergistic effects of the marriage of molecular methods with epidemiology. This powerful combination enabled scientists to follow the emergence and spread, and to identify ways to prevent transmission and further introductions of the virus into human populations.

| Mortality Associated with Influenza pandemics and Selected Seasonal Epidemic Events, 1918–2009.* | | |
|---|---|---|
| **Years** | **Circulating Virus (Genetic Mechanism)** | **Excess Deaths from Any Cause** *no. per 100,000 persons/yr* |
| 1918–1919 | H1N1 (viral introduction), pandemic | 598.0 |
| 1928–1929 | H1N1 (drift) | 96.7 |
| 1934–1936 | H1N1 (drift) | 52.0 |
| 1947–1948 | H1N1 A' (intrasubtypic reassortment) | 8.9 |
| 1951–1953 | H1N1 (intrasubtypic reassortment) | 34.1 |
| 1957–1958 | H2N2 antigenic shift), pandemic | 40.6 |
| 1968–1969 | H3N2 (antigenic shift), pandemic | 16.9 |
| 1972–1973 | H3N2 A Port Chalmers (drift) | 11.8 |
| 1975–1976 | H3N2 (drift) and H1N1 ("swine flu" outbreak) | 11.8 |
| 1977–1978 | H3N2 (drift) and H1N1 ("swine flu" outbreak) | 12.4 |
| 1977–1999 | H3N2 (drift) A Sydney (intrasubtypic reassortment) and H1N1 (drift) | 49.5 |
| 2003–2004 | H3N2 Fujian (intrasubtypic reassortment and H1N1 (drift) | 17.1 |
| 2009 | H3N2 and H1N1 (drift) and swine-origin H1N1 (viral introduction), pandemic | ? |

*Mortality data include deaths associated with all influenza A and B viruses combined. Many of these data have been calculated with the use of differing methods and may not be strictly comparable.[1,2] The 1934, 1951, and 1977 data span 2 years.

**FIGURE 3.2**

Mortality associated with influenza pandemics and selected seasonal epidemic events, 1918 to 2009.
*Source: Reproduced, with permission, from Morens et al.[31]*

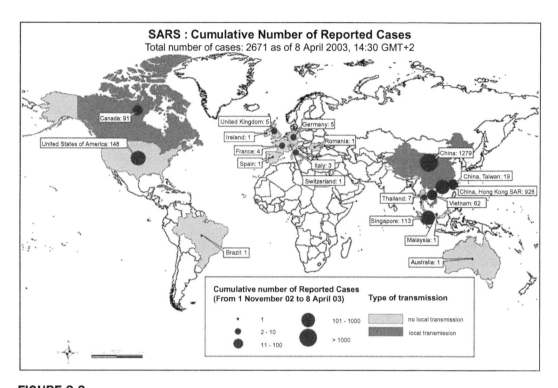

**FIGURE 3.3**

The rapid dissemination of severe acute respiratory syndrome (SARS). *Source: Reproduced, with permission, from the World Health Organization.*[33]

34

SARS was first reported in southern China in 2002 and rapidly spread worldwide (Figure 3.3). Basic epidemiologic methods were essential for tracking the outbreak; a carefully collected epidemiologic case definition was sufficient for case ascertainment, clinical management, infection control, and identification of chains of transmission.[34] However, key to characterizing, and ultimately preventing and controlling the outbreak, was the ability to detect mild cases, and confirm that widely disseminated cases were caused by the same pathogen, which required a validated antibody test.[35] Early in the epidemic there were many possible candidates identified as the cause, but these microbes were not found in all SARS patients. A variety of state-of-the-art and standard molecular techniques were used to identify the viral agent, a new coronavirus. Molecular techniques established that the coronavirus isolated from SARS cases identified worldwide were caused by the same virus, confirming transmission, and a rapidly developed test demonstrated that SARS patients had antibodies to the new coronavirus. That SARS was a virus newly introduced to humans was confirmed by demonstrating that healthy controls not suffering from SARS had no evidence of either past or present infection.[36]

Most microbes have a relatively short life span compared to humans. As they reproduce, mutations occur; further, microbes may exchange genetic material. Therefore, it is not sufficient to compare genetic sequence, because we anticipate there will be changes in the genetic sequence over time. To make sense of genetic changes requires an analysis that takes into account evolutionary relationships, a field called phylogenetics. Phylogenetics enables epidemiologists to trace the emergence and transmission of a rapidly evolving species and, in an outbreak situation, determine order of transmission. HIV, which causes AIDS, evolves quite rapidly even with a single host. Thus the strain that infects an individual is not genetically identical to the strains that the individual might transmit to others. This property of HIV has made it possible to confirm the deliberate infection of one individual by another using a single blood sample from an individual[37] and to gain insight into the spread of HIV worldwide. Not all HIV subtypes spread at the same rate. Using phylogenetic analysis Saad and colleagues[38] traced the introduction and spread of HIV in the Ukraine. The analysis revealed that two HIV subtypes introduced into drug networks in the 1990s still contributed to the epidemic in 2001 and 2002; and that one subtype spread widely throughout the Ukraine and into Russia, Moldova, Georgia, Uzbekistan, and Kyrgyzstan. Further studies to determine the biological and social contributions to the success of the one subtype over another will provide important insights into how to control HIV.

### 3.4.4 Trace Mobile Genetic Elements

Mobile genetic elements are sequences of genetic material that can change places on a chromosome, and be exchanged between chromosomes, between bacteria, and even between species. A type of mobile genetic element known as a plasmid can integrate directly into the chromosome or survive as extrachromosomal material in the cytoplasm of bacteria and code for proteins. The recognition of mobile genetic elements and the ability to trace these genetic elements as they move within and between species has caused a rethinking of the rate of and potential for microbial evolution. For example, Shiga toxin–producing E. coli probably emerged from the transfer of genes coding for Shiga toxin from *Shigella* into E. coli. Antibiotic resistance is often spread via mobile genetic elements, which tend to code for genes providing resistance against multiple antibiotics. This explains several apparent mysteries, such as the spread across many bacterial species within a hospital of the same antibiotic resistance profile, and why treating an individual with one antibiotic can result in resistance to multiple, unrelated antibiotics. Because mobile genetic elements evolve separately from their microbial hosts, separate phylogenies can be constructed, giving insight into the emergence and evolution of these elements.

## 3.5 IDENTIFY PREVIOUSLY UNKNOWN OR UNCULTIVABLE INFECTIOUS MICROBES

The vast majority of microbes cannot be cultured using standard laboratory techniques; the ability to make a copy of genetic material and determine the genetic sequence, which can then be compared to known genetic sequence, has led to a radical reassessment of the amount of life around, in, and on us. Nonculture techniques have enabled us to characterize the microbial communities living in the mouth, gut, vagina, and other body sites, and the detection of microbes in body sites previously thought to be sterile, such as the blood. Epidemiologic data may suggest an infectious origin for a disease; in the past, if an organism could not be cultured, it remained only a suggestion. Molecular tools have changed this by enabling the detection of uncultivable microbes. Two examples of the power of combining epidemiologic principles with molecular techniques are the identification of the causative agent of Kaposi sarcoma, and the identification of HPV as the cause of cervical cancer, which led to the development of an effective vaccine.

The epidemiology of Kaposi sarcoma among persons with AIDS strongly suggested that it was caused by an infectious agent. Not all persons with AIDS had Kaposi sarcoma, and AIDS patients with specific characteristics were more likely to get the sarcoma. Kaposi sarcoma occurred much more frequently among men who have sex with men compared to heterosexual populations or those who acquired HIV from blood or blood products, and among those engaging in specific sexual practices. Reasoning that Kaposi sarcoma might be caused by a virus, Chang and associates[39] used representative difference analysis, a technique that identifies DNA sequences present in one set of tissue but absent in another, to identify and characterize unique DNA sequences in Kaposi sarcoma tissue that were either absent or present in low copy number in nondiseased tissue obtained from the same patient. Chang and colleagues[39] then went on to demonstrate that the sequences, which were similar to herpes virus, were not present in tissue DNA from non-AIDS patients, but were present in

AIDS and non-AIDS patients with Kaposi sarcoma worldwide,[40,41] and that infection with the virus preceded development of the sarcoma.[42,43]

HPV types 16 and 18 are now known to cause cervical and other cancers, and a vaccine is licensed to prevent acquisition. HPV 16 was first identified in 1983 before the virus could be grown.[44] When HPV 16 was discovered, we were aware that papillomavirus could cause cancer in rabbits, cows, and sheep, but it was not clear that the HPV caused cancer in humans. HPV was a suspected cause of genital cancer, because – similar to Kaposi's sarcoma – the epidemiology suggested that an infectious agent was involved. However, other genital infections, particularly herpes simplex virus, were also suspects. HPV had been ruled out by many, but a new molecular technique, the hybridization assay, detected in cancerous tissue a new subtype, HPV 16, which was specifically associated with cervical and other cancers.[44] The association of HPV 16 with cancer was confirmed by comparing presence of HPV 16 among cancer patients to controls. Though this evidence can be very suggestive, it does not demonstrate temporal order, because the cancer might have preceded HPV 16 infection. Demonstrating temporal order required large-scale prospective cohort studies. These studies also provided key insights supporting the possibility that HPV could be prevented by vaccination, because reinfection with the same HPV subtype rarely occurred, and antibody could protect against reinfection and persistence of low-grade lesions.[45]

## 3.6 PROVIDE INSIGHT INTO PATHOGEN GENE FUNCTION AND HOST–PATHOGEN INTERACTION

Many major human pathogens have been genetically sequenced, and hundreds of microbial genomes will be sequenced in the near future. (For a listing, see the website, Pathogen Genomics, at http://www.sanger.ac.uk/Projects/Pathogens.[46]) Unlike genes from multicellular organisms, single-celled organisms often vary greatly in genetic content and expression – that is, genes may be present or absent as well as expressed or silent. Once a single strain of a microbe has been sequenced, the sequence can be used as a reference for comparison with others in the same species, providing insight into the heterogeneity of the species. Sequence information can be mined for potential virulence genes, by identifying genes of unknown function with structures similar to genes whose function is known in the same or other species. However, until the presence or expression of the gene is associated with transmission, pathogenesis, or virulence at a population level, the relative importance of a particular gene cannot be discerned. A gene that is associated with virulence in the laboratory may occur so rarely in vivo that it is an inappropriate target for vaccination or therapeutics. Epidemiologic screening of collections of human pathogens for the prevalence of genes that alter the transmission, pathogenesis, and virulence of the microbe provides insight into the potential importance and putative function of genes identified using genomic analyses. A gene found more frequently in strains that cause severe disease (virulent strains) than among strains that colonize without causing symptoms (commensal strains) suggests that the gene is worthy of more detailed laboratory analyses of its function. A genomic subtraction of a *Haemophilus influenzae* causing middle ear infection (otitis media) from a laboratory strain of *H. influenzae* identified several genes found only in middle ear isolates. Upon screening of collections of human isolates, one gene, *lic*2B, occurred 3.7 times more frequently among middle ear than colonizing isolates, suggesting that *lic*2B is involved in the pathogenesis of otitis media.[47]

Not all genes are expressed at all times; expression profiling (transcriptomics) can identify which genes a pathogenic microbe expresses at different stages of pathogenesis, and which genes the human host expresses in response. Using expression arrays, expression of distinct sets of genes associated with acute, asymptomatic, and the AIDS stages of HIV-1 infection were detected.[48] While only a first step, this type of descriptive analysis provides a basis for understanding the role of these genes in HIV-1 pathogenesis.

Some parasites live in very different environments during their life cycle; understanding which genes are expressed during different points of development is essential for identifying targets for control either via therapeutics or vaccination. The schistosome, a parasite that infects ~200 million people worldwide and causes a variety of adverse health outcomes, at different points lives in freshwater, snails, and vertebrate hosts. Gene expression studies have identified hundreds of genes that are differentially expressed over the life cycle, many that have potential to be targets for intervention.[49] The bacteria *Porphyromonas gingivalis* inhabits the human mouth and is associated with periodontal disease. Smokers have higher rates of periodontal disease and of persistent *P. gingivalis* infection, suggesting that smoking may modify *P. gingivalis* interaction with the host. This was shown to be the case in an experiment that used a microarray representative of the *P. gingivalis* genome to monitor expression. Exposure to cigarette smoke extract changed regulation of a number of *P. gingivalis* genes, and monoctyes and peripheral blood mononuclear cells had a lower proinflammatory response when *P. gingivalis* was exposed to cigarette smoke extract.[50]

# References

1. J.R. Harris, D. Bergmire-Sweat, J.H. Schlegel, et al., Multistate outbreak of *Salmonella* infections associated with small turtle exposure, 2007–2008, Pediatrics 124 (5) (2009) 1388–1394.

2. Centers for Disease Control and Prevention, Division of Tuberculosis Elimination. Guide to the Application of Genotyping to Tuberculosis Prevention and Control. [Internet]. Available at: http://www.cdc.gov/tb/programs/genotyping/Append/AppendixB_4_Core.htm, 2009 (Updated 1.6.09; accessed 5.8.10).

3. Centers for Disease Control and Prevention, Multistate outbreak of human *Salmonella* infections caused by contaminated dry dog food – United States, 2006-2007, MMWR Morb. Mortal. Wkly. Rep. 57 (19) (2008) 521–524.

4. Centers for Disease Control and Prevention: Department of Informatics Shared Services. Public Health Surveillance Slide Set [Internet]. Available at: http://www.cdc.gov/ncphi/disss/nndss/phs/overview.htm, 2009 (Updated 5.1.09; accessed 14.12.09).

5. M. Trostle, USAID: Infectious Diseases: Disease Surveillance: Overview. [Internet]. Washington, DC: United States Agency for International Development. Available at: http://www.usaid.gov/our_work/global_health/id/surveillance/ (accessed 25.2.10).

6. A.G. Mainous, 3rd, W.J. Hueston, C.J. Everett, et al., Nasal carriage of *Staphylococcus aureus* and methicillin-resistant *S aureus* in the United States, 2001–2002, Ann. Fam. Med. 4 (2) (2006) 132–137.

7. P.L. Lu, L.K. Siu, T.C. Chen, et al., Methicillin-resistant *Staphylococcus aureus* and *Acinetobacter baumannii* on computer interface surfaces of hospital wards and association with clinical isolates, BMC Infect. Dis. 9 (2009) 164.

8. B. Swaminathan, T.J. Barrett, S.B. Hunter, et al., PulseNet: The molecular subtyping network for foodborne bacterial disease surveillance, United States, Emerg. Infect. Dis. 7 (3) (2001) 382–389.

9. Centers for Disease Control and Prevention, Ongoing multistate outbreak of *Escherichia coli* serotype O157:H7 infections associated with consumption of fresh spinach – United States, September 2006, MMWR Morb. Mortal. Wkly. Rep. 55 (38) (2006) 1045–1046.

10. V.I. Siarkou, D. Vitti, E. Protonotariou, et al., Molecular epidemiology of outbreak-related pseudomonas aeruginosa strains carrying the novel variant blaVIM-17 metallo-beta-lactamase gene, Antimicrob. Agents Chemother. 53 (4) (2009) 1325–1330.

11. J.E. Golub, W.A. Cronin, O.O. Obasanjo, et al., Transmission of *Mycobacterium tuberculosis* through casual contact with an infectious case, Arch. Intern. Med. 161 (18) (2001) 2254–2258.

12. D. Medini, D. Serruto, J. Parkhill, et al., Microbiology in the post-genomic era, Nat. Rev. Microbiol. 6 (6) (2008) 419–430.

13. M. Kibur, V. af Geijerstamm, E. Pukkala, et al., Attack rates of human papillomavirus type 16 and cervical neoplasia in primiparous women and field trial designs for HPV16 vaccination, Sex Transm. Infect. 76 (1) (2000) 13–17.

14. R. Guy, J. Gold, J.M. Calleja, et al., Accuracy of serological assays for detection of recent infection with HIV and estimation of population incidence: A systematic review, Lancet Infect. Dis. 9 (12) (2009) 747–759.

15. J. Simonsen, K. Molbak, G. Falkenhorst, et al., Estimation of incidences of infectious diseases based on antibody measurements, Stat. Med. 28 (14) (2009) 1882–1995.

16. B. Foxman, B. Gillespie, S.D. Manning, et al., Incidence and duration of group B Streptococcus by serotype among male and female college students living in a single dormitory, Am. J. Epidemiol. 163 (6) (2006) 544–551.

17. C. Gardella, Z. Brown, A. Wald, et al., Risk factors for herpes simplex virus transmission to pregnant women: A couples study, Am. J. Obstet. Gynecol. 193 (6) (2005) 1891–1899.

18. B. Foxman, S.D. Manning, P. Tallman, et al., Uropathogenic *Escherichia coli* are more likely than commensal *E. coli* to be shared between heterosexual sex partners, Am. J. Epidemiol. 156 (12) (2002) 1133–1140.

19. S.D. Manning, K. Neighbors, P.A. Tallman, et al., Prevalence of group B streptococcus colonization and potential for transmission by casual contact in healthy young men and women, Clin. Infect. Dis. 39 (3) (2004) 380–388.

20. A.I. Khan, K.A. Talukder, S. Huq, et al., Detection of intra-familial transmission of shigella infection using conventional serotyping and pulsed-field gel electrophoresis, Epidemiol. Infect. 134 (3) (2006) 605–611.

21. M.E. Kolader, N.H. Dukers, A.K. van der Bij, et al., Molecular epidemiology of *Neisseria gonorrhoeae* in Amsterdam, The Netherlands, shows distinct heterosexual and homosexual networks, J. Clin. Microbiol. 44 (8) (2006) 2689–2697.

22. K. DeRiemer, C.L. Daley, Tuberculosis transmission based on molecular epidemiologic research, Semin. Respir. Crit. Care Med. 25 (3) (2004) 297–306.

23. B. Mathema, N.E. Kurepina, P.J. Bifani, et al., Molecular epidemiology of tuberculosis: current insights, Clin. Microbiol. Rev. 19 (4) (2006) 658–685.

24. D. Lundback, H. Fredlund, T. Berglund, et al., Molecular epidemiology of *Neisseria gonorrhoeae*: identification of the first presumed Swedish transmission chain of an azithromycin-resistant strain, APMIS 114 (1) (2006) 67–71.

25. G.V. Doern, S.S. Richter, A. Miller, et al., Antimicrobial resistance among *Streptococcus pneumoniae* in the United States: have we begun to turn the corner on resistance to certain antimicrobial classes? Clin. Infect. Dis. 41 (2) (2005) 139–148.

26. J. Reidl, K.E. Klose, *Vibrio cholerae* and cholera: out of the water and into the host, FEMS Microbiol. Rev. 26 (2) (2002) 125–139.

27. NNIS System, National Nosocomial Infections Surveillance (NNIS) System Report, data summary from January 1992 through June 2004, issued October 2004, Am. J. Infect. Control 32 (8) (2004) 470–485.

28. T.S. Naimi, K.H. LeDell, K. Como-Sabetti, et al., Comparison of community- and health care-associated methicillin-resistant *Staphylococcus aureus* infection, JAMA 290 (22) (2003) 2976–2984.

29. K.E. Sabol, K.L. Echevarria, J.S. Lewis, 2nd, Community-associated methicillin-resistant Staphylococcus aureus: new bug, old drugs, Annals of Pharmacotherapy 40 (6) (2006) 1125–1133.

30. K. Van Reeth, Avian and swine influenza viruses: our current understanding of the zoonotic risk, Vet. Res. 38 (2) (2007) 243–260.

31. D.M. Morens, J.K. Taubenberger, A.S. Fauci, The persistent legacy of the 1918 influenza virus, N. Engl. J. Med. 361 (3) (2009) 225–229.

32. V.C. Cheng, S.K. Lau, P.C. Woo, et al., Severe acute respiratory syndrome coronavirus as an agent of emerging and reemerging infection, Clin. Microbiol. Rev. 20 (4) (2007) 660–694.

33. World Health Organization. SARS: Cumulative Number of Reported Cases: Total Number of Cases. 2671 as of 8 April, 2003 14:30 GMT+2 [Internet]. Available at: http://www.who.int/csr/sars/SARS2003_4_8.jpg, 2003 (accessed 5.8.10).

34. R.A. Weinstein, Planning for epidemics: the lessons of SARS, N. Engl. J. Med. 350 (23) (2004) 2332–2334.

35. A. Berger, C. Drosten, H.W. Doerr, et al., Severe acute respiratory syndrome (SARS): paradigm of an emerging viral infection, J. Clin. Virol. 29 (1) (2004) 13–22.

36. T.G. Ksiazek, D. Erdman, C.S. Goldsmith, et al., A novel coronavirus associated with severe acute respiratory syndrome, N. Engl. J. Med. 348 (20) (2003) 1953–1966.

37. M.L. Metzker, D.P. Mindell, X.M. Liu, et al., Molecular evidence of HIV-1 transmission in a criminal case, Proc. Natl. Acad. Sci. USA 99 (22) (2002) 14292–14297.

38. M.D. Saad, A.M. Shcherbinskaya, Y. Nadai, et al., Molecular epidemiology of HIV Type 1 in Ukraine: birthplace of an epidemic, AIDS Res. Hum. Retroviruses 22 (8) (2006) 709–714.

39. Y. Chang, E. Cesarman, M.S. Pessin, et al., Identification of herpesvirus-like DNA sequences in AIDS-associated Kaposi's sarcoma, Science 266 (5192) (1994) 1865–1869.

40. P.S. Moore, Y. Chang, Detection of herpesvirus-like DNA sequences in Kaposi's sarcoma in patients with and without HIV infection, N. Engl. J. Med. 332 (18) (1995) 1181–1185.

41. Y. Chang, J. Ziegler, H. Wabinga, et al., Kaposi's sarcoma-associated herpesvirus and Kaposi's sarcoma in Africa. Uganda Kaposi's Sarcoma Study Group, Arch. Intern. Med. 156 (2) (1996) 202–204.

42. S.J. Gao, L. Kingsley, D.R. Hoover, et al., Seroconversion to antibodies against Kaposi's sarcoma-associated herpesvirus-related latent nuclear antigens before the development of Kaposi's sarcoma, N. Engl. J. Med. 335 (4) (1996) 233–241.

43. P.S. Moore, L.A. Kingsley, S.D. Holmberg, et al., Kaposi's sarcoma-associated herpesvirus infection prior to onset of Kaposi's sarcoma, AIDS 10 (2) (1996) 175–180.

44. M. Durst, L. Gissmann, H. Ikenberg, et al., A papillomavirus DNA from a cervical carcinoma and its prevalence in cancer biopsy samples from different geographic regions, Proc. Natl. Acad. Sci. USA 80 (12) (1983) 3812–3815.

45. L. Koutsky, The epidemiology behind the HPV vaccine discovery, Ann. Epidemiol. 19 (4) (2009) 239–244.

46. Wellcome Trust Sanger Institute. Pathogen Genomics [Internet]. Available at: http://www.sanger.ac.uk/Projects/Pathogens/ (accessed 5.8.09).

47. M.M. Pettigrew, B. Foxman, C.F. Marrs, et al., Identification of the lipooligosaccharide biosynthesis gene lic2B as a putative virulence factor in strains of nontypeable *Haemophilus influenzae* that cause otitis media, Infect. Immun. 70 (7) (2002) 3551–3556.

48. Q. Li, A.J. Smith, T.W. Schacker, et al., Microarray analysis of lymphatic tissue reveals stage-specific, gene expression signatures in HIV-1 infection, J. Immunol. 183 (3) (2009) 1975–1982.

49. J.M. Fitzpatrick, E. Peak, S. Perally, et al., Anti-schistosomal intervention targets identified by lifecycle transcriptomic analyses, PLoS Negl. Trop. Dis. 3 (11) (2009) e543.

50. J. Bagaitkar, L.R. Williams, D.E. Renaud, et al., Tobacco-induced alterations to Porphyromonas gingivalis-host interactions, Environ. Microbiol. 11 (5) (2009) 1242–1253.

# CHAPTER 4

# A Primer of Epidemiologic Study Designs

However dazzling the capabilities of modern molecular methods, if they are not applied in properly designed and conducted studies the results will be useless. This maxim is succinctly summarized as GIGO: garbage in, garbage out. The process of designing a molecular epidemiologic study requires balancing the constraints imposed by the molecular tool with study design considerations. These trade-offs are the topic of Chapter 9. The current chapter presents an overview of epidemiologic study designs and an introduction to biases in study design, which can be skipped by the reader already familiar with the topic. This chapter is intended to introduce the vocabulary of epidemiologic study design to facilitate understanding of the examples presented throughout the remainder of the text. The reader is referred to the many epidemiologic textbooks for a more extensive introduction.

Study design dictates what parameters can be estimated and what inferences can be made. A cross-sectional survey that measures a characteristic at a single point in time, such as frequency of antibiotic resistance among those with bacterial infection, cannot give insight into how resistance changes with time. It can be used to describe how antibiotic resistance varies with number of prior courses of antibiotic treatment. The investigator will be uncertain, however, if the resistance led to an increased number of treatments or the increased number of treatments led to the resistance, because individuals were not followed over time. Design choices also include what variables are measured and the population to be studied. These choices also drive what inferences might be made from the results. Resistance mechanisms vary by bacterial species and antibiotic treatments. To whom the results can be generalized depends on which species and therapies are included. Further, the characteristics of the participants are very important. The factors that lead to emergence and spread of antibiotic resistance in an intensive care unit may be quite different from those that occur in the community. Therefore the investigator should have a firm idea of the research question, and what type of information is required to answer it, so that the study can be designed to give the results required to make the desired inferences.

Most epidemiologic study designs are observational. In an observational study, the investigator observes without manipulating the population. An observational study of the effect of hepatitis B virus on risk of liver cancer might identify individuals with hepatitis B virus and watch them over time until they develop liver cancer. In an experiment, individuals would be randomly assigned to be infected with the virus (which of course would not be ethical in humans). The challenge of an observational study is to have the study mimic an experiment, as best as is possible. Individuals are not exposed to hepatitis B virus at random. The factors that lead to viral exposure may be associated with increased risk of liver cancer in the presence of the virus. This makes it more difficult to establish a causal relationship using an observational study design, because the relationship may be explained by factors out of the investigator's control. Therefore observational designs are best at identifying associations.

41

Molecular Tools and Infectious Disease Epidemiology.
© 2012 Elsevier Inc. All rights reserved.

**TABLE 4.1** Characteristics of Epidemiologic Study Designs

| Study Design | Sampling for Inclusion | Timing of Measurement | Purpose |
|---|---|---|---|
| Experiment | By population | Individual: multiple time points | Hypothesis testing |
| Cohort | By exposure or population | Individual: multiple time points | Hypothesis testing |
| Case–control | By disease | Individual: single time point | Hypothesis testing<br>Hypothesis generating<br>Descriptive |
| Cross-sectional | By exposure, outcome, or population | Individual: single time point | Descriptive<br>Hypothesis generating |
| Ecological | By exposure, outcome, or population | Population: single or multiple time points | Hypothesis generating of population-level effects<br>Descriptive |

These associations can be between any two variables, which are generically labeled the exposure and outcome. Exposure does not necessarily mean an individual is literally exposed to a factor; the exposure can be intrinsic, such as having a specific gene allele. Exposure may also be a therapeutic treatment, colonization with a specific microbe, or proximity to a toxic dump. Outcomes can be a disease, a marker of health, or a marker along a biological pathway such as those leading to normal fetal development or cell senescence. An exposure in one study may be an outcome for another and vice versa.

There are four classic observational epidemiologic study designs: cohort, cross-sectional, case–control, and ecological (Table 4.1). These designs can be differentiated by sampling methods; population; outcome; exposure; timing of assessment, at single or multiple time points; and chronological time of data collection. Sampling and timing also determine the parameters that can be estimated (Table 4.2). Participants can be identified and studied in real time (prospectively), using records or stored specimens (retrospectively) or a combination of prospective and retrospective data collection. Epidemiologists also conduct experiments in human populations, usually called clinical or therapeutic trials. However, when the laboratory is fully integrated into the design and conduct of a study, many experiments must be conducted to evaluate the laboratory assessments. These assessments include evaluations of the validity and reliability of any measuring instruments, the sensitivity of proposed assays to proposed protocols for specimen collection, transport and storage, and the safety of procedures.

Each study design is described, with examples, in the remainder of the chapter. The chapter closes with a brief overview of biases that can occur during the process of study design.

## 4.1 EXPERIMENT

An experiment is the most powerful study design, because the investigator assigns the exposure or treatment. Exposure assignment enables the investigator to make the comparison groups as similar as possible, so that any differences observed should be attributable to the exposure or treatment. Ideally, experimental subjects will be assigned to the treatment at random, and the investigator will be masked to the treatment when assessing outcome, to avoid any biases in assignment or assessment.

In molecular epidemiology, there are two major applications of the experimental design. The first is in the design and conduct of clinical trials in humans. Clinical trials are used to assess the efficacy of vaccines, drugs, or other interventions. There is an extensive literature

**TABLE 4.2 Parameters That Can Be Estimated from Selected Epidemiologic Study Designs**

| Study Design | Parameters That Can Be Estimated | Calculation |
|---|---|---|
| Experiment | Incidence Rate (density) (IR) | New cases/person-time |
| | Cumulative incidence (risk) (CI) | New cases/population at risk |
| | Rate ratio (RR) | IR treated/IR untreated |
| | Cumulative incidence ratio (CIR or risk ratio) | CI treated/CI untreated |
| | Odds Ratio (OR) (tends to exaggerate the CIR when incidence is high) | Odds of disease in treated/odds of disease in untreated |
| | Population attributable risk or rate (PAR) | IR (or CI) in population – IR (or CI) in unexposed |
| | Attributable risk or rate (AR) | IR (or CI) in the exposed – IR (or CI) in the unexposed |
| | Correlation coefficient | |
| Cohort | Prevalence of disease | No. with disease/No. sample |
| | Prevalence of exposure | No. with exposure/No. sample |
| | IR | |
| | CI | |
| | RR | |
| | CIR | |
| | OR (tends to exaggerate the CIR when incidence is high) | |
| | PAR | |
| | AR | |
| | Correlation coefficient | |
| Cross sectional | Prevalence of disease | Prevalence of disease = Incidence × Duration × (1-Prevalence of disease) |
| | Prevalence of exposure | |
| | Prevalence odds ratio (POR) | |
| | Exposure odds ratio (EOR) | |
| | If population is fixed and exposure does not change with time: incidence and risk ratio | |
| | Point Prevalence Ratio: If duration does not vary with exposure and disease is rare approximates the risk ratio | Prevalence in exposure/Prevalence in unexposed = Risk Ratio × (Duration of exposed/Duration unexposed) × (1-Prevalence exposed)/(1-Prevalence unexposed) |
| | Correlation coefficient | |
| Case-Control | (1) if sample from total cohort at baseline (case-cohort) | |
| | Exposure OR = Cumulative Incidence (attack rates) | |
| | Population Attributable Risk | (Prevalence of exposure among controls × (EOR-1))/(Prevalence of exposure among controls × ((EOR -1) +1) |
| | Attributable Proportion among exposed | (EOR – 1)/EOR |
| | (2) if sample from individuals at risk when cases occur during follow-up (Risk set or density sampling) | |
| | EOR= rate ratio | |
| | Rate ratio=CIR if rare disease and representative controls | |
| | Attributable proportion among exposed | |
| | PAR if know Pe in pop and rare disease | (Prevalence exposure in population × (EOR-1))/ (Prevalence exposure in population × ((EOR -1) +1) |
| | (3) if sample for survivors OR= OR in population = CIR when disease is rare | |
| Ecologic | Correlation coefficient | |

on the design, conduct, and analysis of clinical trials (see for example Meinert and Tonascia[1] [1986]). Molecular tools can enhance the value of the trial by providing an assessment of compliance with treatment and detecting markers of outcome. The second and more common application is to conduct experiments related to planning and implementing the use of a molecular measure in an epidemiologic context. Experiments are conducted to optimize laboratory assays for the proposed application; to assess the reliability and validity of molecular measures; to determine the effects of different storage and handling conditions on the variability of study results; to assess the effects of changes in a test kit, reagent, or equipment; and to assess the safety of procedures. These experiments often are not published in the scientific literature or are only summarized briefly in the methods section. However, these experiments are essential to appropriately conducting a study. Therefore the design and conduct of these experiments should be held to the same high standards as other components of the study. This means random selection of specimens for inclusion, random assignment to treatment, and masking of those assessing the effect of the treatment.

Avian influenza virus requires a higher level of biosafety precautions than usually found in a clinical laboratory. The tests used to describe the incidence and prevalence in humans detect antibody to the virus. This requires using live virus and thus currently must be done in special laboratories. Garcia and colleagues[2] optimized and evaluated a diagnostic test for avian influenza that can be conducted safely in a clinical diagnostic laboratory. The test uses a pseudoparticle produced using synthetic genes based on the sequence of the virus as the antigen to detect antibody response. Optimization included determining the optimal amount of the pseudoparticle to use in the assay to detect antibody, and the optimal cutpoint. They then evaluated the reproducibility of the assay over three batches of three plates, each containing six replicates. The batches were tested in three different weeks. To validate the assay, the antibody titers measured using the pseudoparticle were compared to the titers obtained using avian influenza virus. They estimated the sensitivity, specificity, and predictive value of the assay using a panel of control and positive sera. Finally, they checked for the specificity of the assay to avian influenza by testing against reference sera immune to other influenza virus types.

Mayo and Beckwith[3] conducted a series of experiments to assess the safety of procedures used for processing and handling West Nile virus (WNV) (see Chapter 12 for an overview of biosafety). WNV is considered a class 3 agent; regulations for class 3 agents require special laboratory precautions, including physical containment. These precautions are not available in clinical laboratories; however, clinical laboratories test for the presence of antibodies to WNV in body fluids that may also contain WNV. This study assessed whether protocols for testing human serum or cerebrospinal fluid from patients potentially infected with WNV inactivate any WNV present. The particular concern was with an enzyme-linked immunosorbent assay (ELISA) for WNV antibodies. The ELISA involved a washing step that might aerosolize WNV. The concern was that the aerosolized WNV might be sufficient to infect laboratory personnel. The investigators used two levels of WNV in their experiments, one that was several thousand times higher than might be expected in human specimens. They varied the time of the wash and used a plaque assay to detect the presence of WNV. Plaque-forming units are a measure of the infectious virus particles present. After 30 or more minutes in wash buffer containing 0.05% Tween 20, no virus could be detected at either level of WNV. This suggested that by using a longer wash step, the assay might be safely used in a clinical laboratory.

## 4.2 COHORT STUDY

A cohort study attempts to emulate an experiment as much as is possible, given that the investigator has no control over exposure. Cohort studies are used to test and generate hypotheses about causal relationships: does this exposure cause or prevent this outcome? Individuals are sampled for inclusion by exposure status. If exposure status is not known a priori, a defined population is sampled and exposure status determined. The time and expense of mounting a prospective study generally leads to measuring multiple exposures and outcomes, although the study may be powered only to test the primary study hypotheses.

A hallmark of a cohort study is that individuals are followed over time. Observation may occur entirely in the past (a retrospective cohort), partially in the past, or entirely in the present (prospective cohort). Because individuals are followed over time, the investigator can determine that exposure preceded the outcome of interest, and can estimate the number of new cases (incident cases) of outcome that occur. This makes a cohort study the strongest observational study design for assessment of causality. Following participants over time means that it is possible to estimate the dynamics of the underlying system of interest.

Like experimental designs, cohort studies enable estimation of the risk (probability) and rate (occurrence/population/time) of the outcome of interest, and the relative risk and relative rate of outcome by exposure status (Table 4.2). Similar to a clinical trial, molecular tools can be used to assess early markers of disease, potentially decreasing length of follow-up. This decreases costs, enabling enrollment of a larger sample size, which increases the power of the study to detect an effect. Molecular tools also provide more sensitive and specific measures of exposure, thereby decreasing misclassification and increasing study power. Finally molecular tools provide different types of data than are classically included in an epidemiologic study. For example, gene variants might be used as an outcome, or microbial genes with specific functions from a metagenome analysis could be used as exposure.

Molecular tools can greatly increase the value of already collected cohort data. A well-conducted cohort with stored samples can be revisited to examine the association between exposures and the study outcome that may not have been thought of at the time the study was conducted. Similarly, the investigator may be able to identify early markers of other diseases in stored specimens, enabling the investigator to reanalyze the cohort to examine a completely different set of outcomes. The utility of using stored samples and a cohort design are illustrated by a study of herpes simplex virus 1 (HSV1) transmission.[4] By adulthood, most individuals in the United States have been exposed to and infected with HSV1. Those infected with HSV1 may be completely asymptomatic. However, others have mild symptoms such as mild oral or genital lesions, or suffer from significant disease, such as meningitis or encephalitis.[4] Although cross-sectional studies have established that most HSV1 is acquired during childhood, only a study with a follow-up component could establish when HSV1 is generally acquired and from whom. Tunback and associates[4] conducted a retrospective

45

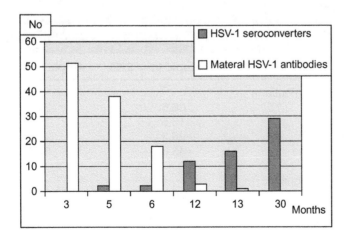

**FIGURE 4.1**
In a study of stored blood samples from 129 infant–mother pairs, 65% of mothers were HSV1-positive, and 68% of their infants had maternal antibodies. By 30 months, infants of seropoitive mothers were much more likely than infants of seronegative mothers to seroconvert (33% vs. 16%). *Source: Reproduced, with permission, from Tunback et al.[4] (2007).*

cohort study using blood samples collected from 129 Swedish infants and mothers as part of a vaccine trial conducted in the early 1990s (Figure 4.1). Sixty-five percent of the mothers had HSV1 antibodies; 68% of infants born of a seropositive mother had detectable maternal antibodies. No infants of seronegative mothers had detectable maternal antibodies. Maternal antibodies were mostly undetectable by the time an infant attained 13 months of age; and by 30 months, 33% of children born to seropositive mothers compared to 16% of children born to seronegative mothers had seroconverted. These data suggest that much of HSV1 transmission occurs during infancy. Because day care is unusual in this study population before age 12 months, it further suggests that most HSV1 transmission in this population occurs within families.

## 4.3 CROSS-SECTIONAL STUDY

A cross-sectional study samples by exposure, outcome, or population, and participants are measured at a single point in time. Cross-sectional surveys are used to estimate the frequency of exposure and outcome in a defined population (Table 4.2). Cross-sectional surveys answer such questions as: How often does a bacterial antigen identified as a potential vaccine candidate occur among bacterial strains isolated from the general population? What is the variability of microbial community structure between individuals? Does microbial community structure vary by age? Gender? How frequently do bacterial isolates causing disease carry a putative virulence gene compared to commensal isolates of the same species?

Cross-sectional studies also are very useful for generating hypotheses. These studies can be analyzed to examine correlations between existing cases and existing exposures. A limitation of this type of analysis is that existing cases tend to be those who have had disease longer, and those with a rapid course may be missed entirely. Nonetheless, these types of analyses can be very useful, answering such questions as: Is carriage of antibiotic-resistant organisms found more frequently among individuals reporting a recent history of antibiotic use? Are prevalent cases more likely than controls to have positive serology for certain microbes? Is the metabolic profile of gut microbiota among those with the outcome different from those without the outcome?

If specimens are collected, molecular tools can greatly enhance the richness as well as accuracy of data, and the potential analyses of a cross-sectional study. A cohort study implies that individuals are followed over time, but if exposure is something that doesn't change

**TABLE 4.3** Prevalence of Positive Serology for Fecal–Oral Food-Borne and Contact–Airborne Infections Among Persons with and without Atopia in Iceland, Estonia, and Sweden

| | Nonatopic (n = 922) | Atopic (n = 327) | P Value |
|---|---|---|---|
| Fecal–oral/food-borne infections | | | |
| • Helicobacter pylori | 36.6 | 24.8 | <.001 |
| • Hepatitis A virus | 13.6 | 15.8 | .32 |
| • Toxoplasmosis gondii | 26.2 | 22.6 | .21 |
| No fecal–oral/food-borne infections | 45.6 | 54.1 | <.008 |
| Contact/airborne infections | | | |
| • HSV1 | 76.7 | 67.4 | .001 |
| • EBV | 90.9 | 88.1 | .15 |
| • Chlamydia pneumoniae | 68 | 58.8 | .002 |
| • Cytomegalovirus | 81 | 74.6 | .01 |
| Two or fewer contact/airborne infections | 21.1 | 31.4 | <.001 |
| ≤3 all infections | 34.8 | 51.7 | <.001 |

Source: Adapted, with permission, from Janson et al.[5] (2007).

46

with time, such as genetic makeup, the results of a cross-sectional study can be analyzed as a cohort study. This is straightforward for diseases with a short etiological period (time from exposure to development of detectable outcome), where there is little potential of loss to the cohort due to a combination of exposure and outcome. For diseases with a longer etiological period, analyzing a cross-sectional study as a cohort may still be informative, but the investigator must take into account any potential losses to the cohort that may have happened over time that could bias study results. If those with a factor under study are more likely to die early of the outcome of interest, the results might suggest an erroneous beneficial association between exposure and outcome. To create an extreme example, suppose 30% of a cohort is exposed, and 10% exposed develop the outcome compared to 5% of those not exposed. If 10% of the exposed population dies before the study takes place, and all of those are the exposed with the outcome, none of the exposed would have disease, so the exposure would appear protective. However, for acute exposures and outcomes where the timing can be determined, the cohort analysis is more powerful.

Janson and associates[5] (2007) used a cross-sectional design to test a prediction of the hygiene hypothesis. The hygiene hypothesis posits that appropriate exposure to microbes early in life is required to set up immunoregulatory pathways. In the absence of this early exposure to microbes, the body may have a hyperactive response to allergens in the environment. This hyperactive response may lead to an increase in atopic diseases, such as eczema and asthma. In order to test the hypothesis, Janson and colleagues[5] measured the prevalence of immunoglobulin E (IgE) against *Dermatophagoides pteronyssinus*, timothy grass, cat, and *Cladosporium herbarum*, and immunoglobulin G (IgG) against *Helicobacter pylori*, *Toxoplasmosis gondii*, hepatitis A virus, herpes simplex virus 1, *Chlamydia pneumoniae*, Epstein Barr virus (EBV), and cytomegalovirus in 1249 persons in Iceland, Estonia, and Sweden.[5] They compared the prevalence of antibodies to the various infections among those with and without atopy, defined as sensitivity to any of the allergens measured using IgE. The comparison of frequency of exposure (antibodies) among those with and without outcome (atopy) is the same strategy used to analyze a case–control study (see Section 4.4). For many of the infections measured, the prevalence of antibodies was lower among those with atopia, supporting the hygiene hypothesis (Table 4.3). Further, those with fewer infections, regardless of type, were more likely to be atopic.

## 4.4 CASE–CONTROL STUDY

Sampling for a case–control study is by outcome, which is generally a disease (called a case). Case–control studies are used to generate and test hypotheses about associations between exposures and outcomes. Because sampling is by outcome, case–control studies are optimal for rare diseases where following a cohort would be prohibitive. The case–control design also lends itself to secondary analyses of data from large cohort studies conducted for another reason.

Participants may be sampled from records (retrospective) or collected as cases occur (prospective) or a mixture of the two. The parameters that can be estimated depend on how controls are selected (Table 4.2). Controls can be sampled from the population at risk at the beginning of follow-up (cohort or base sample), from the population at risk at the end of follow-up who did not develop of the disease (survivor sampling), or from the population at risk as cases are diagnosed (risk set sampling).[6] Regardless of the sampling scheme, controls should be sampled from the same underlying cohort in which cases occurred to yield a valid comparison. Moreover, regardless of which control group is selected, the measurement of cases and controls should occur at the same time point, so exposure among cases can be determined before disease occurs (cohort sample), at the end of follow-up (survivor sampling), or as disease occurs (risk set sampling). Therefore, if conducting a case–control study nested in a cohort study, the choice of controls may be fixed by how specimens were collected.[7] A corollary is that the type of control sampling may be driven by ability to collect the desired specimens.

**TABLE 4.4** Association of *Trichomonas vaginalis* and Prostate Cancer by Aspirin Use Since Age 20 Years. Men Participating in the Health Professionals Follow-Up Study 1993 to 2000

| | Seronegative | | Seropositive | | |
| --- | --- | --- | --- | --- | --- |
| | Cases/ Controls | OR (95% CI) | Cases/ Controls | OR (95% CI) | Multivariable-Adjusted OR (95% CI)*,† |
| Total prostate cancer | 604/626 | 1.00 | 87/65 | 1.41 (0.99–2.00) | 1.43 (1.00–2.03) |
| Organ-confined ($\leq T_2$ and $N_0M_0$ | 440/455 | 1.00 | 67/52 | 1.35 (0.91–2.00) | 1.36 (0.92–2.02) |
| Low-grade (Gleason sum <7) | 333/342 | 1.00 | 47/38 | 1.30 (0.81–2.09) | 1.27 (0.79–2.06) |
| High-grade (Gleason sum ≥7) | 210/222 | 1.00 | 33/21 | 1.67 (0.93–2.99) | 1.76 (0.97–3.18) |

Source: Adapted, with permission, from Sutcliffe et al.[8] (2006).
NOTE: Too few participants were diagnosed with advanced-stage prostate cancer (n = 62) to investigate its association with T. vaginalis antibody serostatus.
*Estimated by conditional logistic regression.
†Adjusted for race (white, nonwhite) and cumulative family history of prostate cancer through 1996 (yes/no).

It is easiest to obtain controls representing the underlying cohort when a case–control study is nested within an existing cohort study or when cases and controls are selected from a defined population base (such as a health maintenance organization with patient lists). Otherwise it is harder to verify that the controls were selected from the same underlying cohort from which cases arose, which is an important indicator of internal validity. Consider a situation where cases are identified from all those treated at a given hospital and controls selected from those in the same catchment area. If there are two hospitals with the same catchment area, the choice of hospital might depend on a variety of factors that may or may not be associated with the exposures of interest. This would suggest that the controls might not represent the same underlying cohort as cases. Control selection is closely scrutinized when case–control studies are evaluated for funding and publication in the scientific literature.

Case–control studies are extremely common, both because of the ease of conduct and limited cost. If nested within a cohort study, investigators can estimate incidence, risk, and rate ratios. There are several large ongoing cohort studies; one is the Health Professionals Follow-up Study. This cohort of 51,529 health professionals enrolled in 1986, and 18,018 of these gave blood samples. This group has been revisited using the case–cohort design to address a variety of different conditions. Because of the large size of the cohort, it is possible to explore various hypotheses regarding the etiologies of relatively rare conditions using a nested case–cohort design. One such study by Sutcliffe and colleagues[8] (2006) explored the association between a sexually transmitted protozoan infection, *Trichomonas vaginalis,* and prostate cancer. Sexually transmitted infections have been hypothesized to increase risk of prostate cancer; the proposed etiological pathway is via inflammation. *T. vaginalis* infection is known to be inflammatory, yet it is often asymptomatic, particularly in men. Cases and controls were selected from study participants who participated in the enrollment blood draw of the Health Professionals Follow-up Study and were free from prostate cancer at that time. Controls were further limited to those without cancer and with a negative prostate-specific antigen (PSA) test, a screening test for prostate cancer.[8] Stored bloods were tested for antibodies to *T. vaginalis* using an ELISA. The 691 cases with prostate cancer were matched to controls by age, the timing of blood draw (year, time of day, and season), and history of PSA testing before the blood draw. More cases (13%) than controls (9%) were seropositive for *T. vaginalis.* The association between prostate cancer and *T. vaginalis* was strongest among those who reported infrequent aspirin use since age 20 years, where there was a twofold increase in risk, after adjustment for other factors (Table 4.4). This finding also suggests that inflammation is an important factor in prostate cancer development, because aspirin is an anti-inflammatory.

**TABLE 4.5** Diagnostic Performance of Two Rapid Malaria Diagnostic Tests and Microscopy Using Real-Time PCR as the Gold Standard for Detection of Monoinfection of *Plasmodium falciparum* Malaria among 308 Febrile Patients in Madagascar

| Test | Sensitivity (95% CI) | Specificity (95% CI) | Positive Predictive Value (95% CI) | Negative Predictive Value (95% CI) |
|------|---------------------|---------------------|-----------------------------------|-----------------------------------|
| Microscopy | 98.8 (94.4–99.9) | 100.0 (98.5–100.0) | 100.0 (96.6–100.0) | 99.5 (97.6–99.9) |
| PALUTOP+4 test (ALL. DIAG, Strasbourg, France, www.alldiag.com) | 95.4 (89.3–98.5) | 97.1 (94.0–98.8) | 93.3 (86.5–97.2) | 98.0 (95.3–99.9) |
| OptiMAL-IT test test (DiaMed AG, Cressier sur Morat, Switzerland, www.diamed.com) | 75.8 (66.1–85.0) | 99.0 (96.8–99.8) | 97.1 (90.6–99.5) | 90.6 (86.3–93.9) |

*Source: Adapted, with permission, from Rakotonirina et al.[9] (2008).*

The case–control design is often used to determine the validity and reliability of a laboratory test. For validity assessments, the cases and controls are samples that are positive and negative by the reference standard; the exposure is assessment by the measure of interest. Similarly, for measures of reliability, the cases and controls are specimens from across the range of values measured by the test, and exposure is the repeated measurement by the same test. Rakotonirina and associates[9] (2008) compared the sensitivity, specificity, and predictive values of two rapid malaria diagnostic tests and microscopy relative to real-time PCR, which is considered the gold standard (Table 4.5). Rapid diagnostic tests are fast, easy to perform, and do not require electricity or specific equipment, as compared to microscopy, which although inexpensive, is labor-intensive, time-consuming, and requires well-trained personnel to differentiate between the species. In this study,[9] microscopy performed best. The probability that microscopy identified *Plasmodium falciparum*, given that it was present (sensitivity), was 98.8%, and the probability that microscopy showed no *P. falciparum*, given that it was absent (specificity), was 100.0%. The prevalence of *P. falciparum* in the study population was 27.8% (87/308); microscopy gave excellent positive and negative predictive values.

## 4.5 ECOLOGIC STUDY

An ecological study compares groups, and allows the investigator to make inferences about how an outcome changes over place or time. Ecological studies often generate hypotheses. The hallmark of an ecological study is that either the exposure or outcome variable is measured in aggregate. Aggregate measures may be molecular in nature, such as measurements of air, soil, water, or foodstuffs. For some infectious diseases, group-level variables may be more informative than measures of individual characteristics. Koopman and colleagues[10] used ecological variables to identify predictors of dengue transmission in 70 localities in Mexico. Ecological variables included temperature, proportion of houses with mosquito larvae on the premises, and the proportion of houses with uncovered water containers. Each of these variables was associated with infection frequency, suggesting that control efforts should be directed at the community level.

Werber and co-workers[11] (2008) used an ecological study design to explore whether Shiga toxin–producing *Escherichia coli* (STEC) found in foods lead to human disease. Serogroups from food isolates were compared to serogroups from patients. Two thirds (72%) of STEC serogroups found in foods were detected in patient isolates. The distribution of serogroups in food was significantly different from that found in patients (Figure 4.2). Only 41 (67%) of the 61 serotypes were found in both food and patients; further, these 41 accounted for 9% of patient isolates and 45% of food isolates. The three serotypes most frequently

49

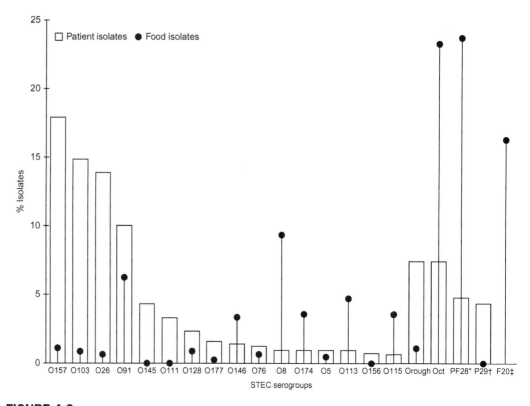

**FIGURE 4.2**

Distributions of Shiga toxin–producing *Escherichia coli* found among isolates from patients (bars) and food (dots) in Germany. *Source: Reproduced, with permission, from Werber et al.*[11] *(2008).*

found in patients accounted for 46% of patient and 3% of food isolates. Although the study demonstrated that serogroups causing human disease were found in food, the two distributions were quite different. The differences in serogroups might reflect true differences in distribution, food preparation among the food sampled that might limit human exposure, sampling schemes, and propensities to cause disease. The variation might also reflect additional transmission routes.

## 4.6 BIASES

In design, conduct, and analysis of epidemiologic studies, we are concerned about three types of bias: misclassification (also known as information bias), selection, and confounding bias. Misclassification bias is discussed in detail in Chapter 2. Briefly, misclassification occurs when individuals are placed into the wrong group, for example, diseased instead of nondiseased and vice versa, usually because of measurement error. No measure is perfect, so some individuals will always be put into the wrong category (misclassified). If wrong assignment occurs at random, the effect is to decrease the association between exposure and outcome. Of greater concern is if the assignment is not at random. Imagine a laboratory technician is testing for an exposure hypothesized to be associated with disease. If not masked to disease status, the technician might be more likely to call questionable results positive for cases and negative for controls. Over the course of the study, this will lead to more cases recorded as having the exposure of interest and fewer controls, even if there is no real association between the exposure and outcome.

Selection bias occurs when study participants are included in the study differentially with respect to outcome and exposure. This does not mean that all groups must have the same probability of inclusion. In a case–control study, participants are sampled by outcome. The

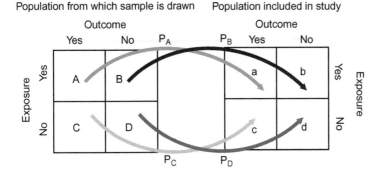

**FIGURE 4.3**
Schematic showing selection from the population of interest for study. If the probability of selection is higher for one group, the estimates of the parameters of interest will be biased (see text).

probability of inclusion is generally higher for cases than controls. This will not result in a bias, unless the probability of selection is different for one case–exposure combination. Consider a population that is classified by exposure and outcome (Figure 4.3). Individuals are sampled for inclusion in the study; the probability of inclusion may vary by cell. Individuals who have both the outcome and exposure (in cell A) have a probability ($P_A$) of selection for inclusion into cell A. If those in cell A have a higher probability of inclusion than those in the other cells, the prevalence of disease among those with exposure will be overestimated. Selection bias might occur in a case–control study if individuals with the outcome and exposure are more likely to participate. Selection bias might occur in a cross-sectional study for the same reason. In a cohort study, selection bias might occur for the same reason, but also if one group is more likely to be followed than others. A selection bias can bias an association away from or toward the null hypothesis of no difference.

Selection biases can be difficult to measure or assess. To prevent selection bias, the investigator should have inclusion and exclusion criteria that are the same for those in all groups. Use of incentives can enhance participation. Analyzing reasons for refusal can help identify if a selection might have occurred. The reasons might also be used to change the protocol to enhance participation. Similarly, there should be protocols in place so follow-up is uniform across all groups. Obtaining permission at time of enrollment to review medical records for cohort studies will enable to investigator to assess the rate of outcome among those lost to follow-up.

Confounding bias occurs when a third variable associated with exposure and outcome causes a distortion of the observed association between exposure and outcome. The direction of bias can be either away from or toward the null hypothesis of no difference. The distortion does not appear when the association is estimated in those with and without the distorting variable; it is only when the groups are combined that the bias occurs. Unless the confounder is measured, it is impossible to assess if a confounding bias occurred. The best prevention is to create a conceptual model of the research program, and think through what variables might be associated both with the exposure(s) and outcome of interest, so they can be measured, and included in the analysis.

## References

1. C.L. Meinert, S. Tonascia, Clinical Trials: Design, Conduct, and Analysis. Monographs in Epidemiology and Biostatistics, vol. 8, Oxford University Press, New York, 1986.

2. J.-M. Garcia, N. Lagarde, E.S.K. Ma, et al., Optimization and evaluation of an influenza A (H5) pseudotyped lentiviral particle-based serological assay, J. Clin. Virol. 47 (1) (2010) 29–33.

3. D.R. Mayo, W.H. Beckwith, 3rd, Inactivation of West Nile virus during serologic testing and transport, J. Clin. Microbiol. 40 (8) (2002) 3044–3046.

4. P. Tunback, T. Bergstrom, B.A. Claesson, et al., Early acquisition of herpes simplex virus type 1 antibodies in children–a longitudinal serological study, J. Clin. Virol. 40 (1) (2007) 26–30.

5. C. Janson, H. Asbjornsdottir, A. Birgisdottir, et al., The effect of infectious burden on the prevalence of atopy and respiratory allergies in Iceland, Estonia, and Sweden, J. Allergy Clin. Immunol. 120 (3) (2007) 673–679.

6. A. Aschengrau, G.R. Seage, Essentials of Epidemiology in Public Health, second ed., Jones and Bartlett Publishers, Sudbury, 2008.

7. A.G. Rundle, P. Vineis, H. Ahsan, Design options for molecular epidemiology research within cohort studies, Cancer Epidemiol. Biomarkers Prev. 14 (8) (2005) 1899–1907.

8. S. Sutcliffe, E. Giovannucci, J.F. Alderete, et al., Plasma antibodies against *Trichomonas vaginalis* and subsequent risk of prostate cancer, Cancer Epidemiol. Biomarkers Prev. 15 (5) (2006) 939–945.

9. H. Rakotonirina, C. Barnadas, R. Raherijafy, et al., Accuracy and reliability of malaria diagnostic techniques for guiding febrile outpatient treatment in malaria-endemic countries, Am. J. Trop. Med. Hyg. 78 (2) (2008) 217–221.

10. J.S. Koopman, D.R. Prevots, M.A. Vaca Marin, et al., Determinants and predictors of dengue infection in Mexico, Am. J. Epidemiol. 133 (11) (1991) 1168–1178.

11. D. Werber, L. Beutin, R. Pichner, et al., Shiga toxin-producing *Escherichia coli* serogroups in food and patients, Germany, Emerg. Infect. Dis. 14 (11) (2008) 1803–1806.

# CHAPTER 5

# A Primer of Molecular Biology

Modern molecular epidemiology is predicated on the use of techniques developed by molecular biologists. Molecular biology is the study of biology at the molecular level, which translates to studying DNA, RNA, proteins, and metabolites of biological processes. Molecular tools enable testing of these substrates in human populations, enhancing and refining the discriminatory ability of epidemiologic studies. When designing a study, epidemiologists need to identify the most informative substrate(s) for answering the research question, balanced with the relative costs, reliability, validity, and acceptability and feasibility of collecting the necessary specimens. The ability to choose the correct measuring instrument for the research question at hand is critical for conducting molecular epidemiologic studies. To make an informed choice requires a basic understanding of molecular biology. This chapter presents a brief introduction intended to give the reader with a limited biology background this basic understanding. The second part of the chapter describes standard laboratory techniques used in molecular biology, and referred to throughout the remainder of the text. Readers familiar with basic molecular biology can continue on to Chapter 6 without any loss of continuity.

## 5.1 CENTRAL DOGMA AND SOME CAVEATS

*DeoxyriboNucleic Acid* (DNA) is double-stranded, with each strand consisting of four nucleotides, adenine (A), tyrosine (T), guanine (G) and cytosine (C). The two strands are held together by hydrogen bonds. The order of the nucleotide bases provides the code for amino acids that make up proteins. The two strands are not identical but made up of complementary nucleotides: A and T bind together with two hydrogen bonds and G and C with three hydrogen bonds. Because of the difference in hydrogen bonds, the GC content of DNA determines the melting temperature needed to separate the two strands. One end of a strand terminates with the fifth carbon on the sugar ring of deoxyribose, known as the 5′ (five prime) end; the other end terminates with the hydroxyl group of the third carbon of deoxyribose, known as the 3′ (three prime) or "tail" end. DNA is synthesized in vivo by reading from the 5′ to 3′ direction. The two strands fit head to tail.

The central dogma of genetics states that information flows from DNA → RNA → protein, and not in the reverse direction. DNA cannot be directly translated to protein; it requires transcription to *RiboNucleic Acid* (RNA), which is then translated to protein. The process of transcription and translation in a eukaryotic cell is shown in Figure 5.1A. The DNA is transcribed into messenger RNA (mRNA) which is translated to protein. An enzyme, RNA polymerase, helps construct mRNA from the DNA template, so mRNA includes the code for the protein and the instructions for protein synthesis. When mRNA is transcribed from DNA for a specific gene coding region, the nucleotide uracil is substituted for thymine. Any introns (intragenetic regions) are excised and the message is brought to the ribosome for

Molecular Tools and Infectious Disease Epidemiology.
© 2012 Elsevier Inc. All rights reserved.

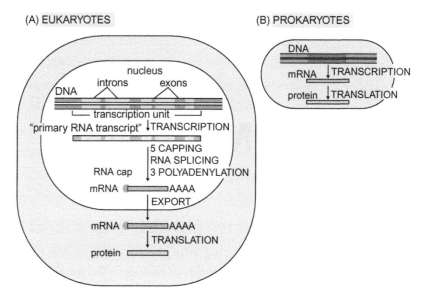

**FIGURE 5.1**

Transcription and translation in eukaryotic and prokaryotic cells. Prokaryotic DNA has little or no intragenetic regions (introns), and prokaryotes do not have a nucleus. *Source: Reproduced, with permission, from Alberts et al.[1] (2004).*

translation to protein. The ribosome translates the mRNA into protein. The protein is not active until it has the correct conformation; and further processing may be required, such as binding to an effector molecule. The process of transcription and translation is similar for prokaryotes (Figure 5.1B). However, prokaryotes lack a nuclear membrane so transcription and translation take place in the cytoplasm. Because introns rarely occur in prokaryotes the mRNA is very similar to DNA, and limited or no posttranscriptional processing is required.

Viruses can either have DNA or RNA as their genetic code; because they have no cells, viruses take over a cell's machinery in order to transcribe and translate their genetic code. DNA viruses follow the central dogma. For some RNA virus, the RNA is directly translated to protein, and in others there is an intermediate step to translate it to mRNA that is translated to protein. Retroviruses are RNA viruses that have a DNA stage; the information flow for a retrovirus is RNA → DNA → RNA → Protein. HIV is a retrovirus. Prions are infectious agents that do not have DNA or RNA; they are made up of protein that is folded in an abnormal form that can propagate the abnormal structure, leading to disease.

## 5.2 MATERIAL TESTED USING MOLECULAR TOOLS

Most of the examples in this book refer to molecular tools that measure DNA, RNA, and proteins in either microbes or humans. However, as we better understand the biological mechanisms behind disease processes, there will be an increasing role for measures that detect changes in cell processes associated with normal or abnormal function (metabolites).

Tests of DNA, the genetic code, look for genetic variations that can uniquely identify an organism, or variations that are associated with health and disease (Table 5.1). The genetic code that results in an individual having a particular characteristic is known as the "genotype." Among bacteria, genes can be present or absent; in addition, there can be variations in gene sequence, copy number, and physical location. In mammalian systems the same set of genes are always present and in the same physical location, but vary in sequence, copy number, or in the genetic material found in between genes (introns). Molecular epidemiology uses the genetic sequence of microbes and their hosts to describe disease patterns as well as to gain insight into possible gene function and origin by describing the distribution of genes or gene variants by person, place, and time characteristics.

**TABLE 5.1** Material Tested Using Molecular Tools, and Why

| Material Tested | What We Are Looking for |
|---|---|
| DNA | Genes, gene variants, genetic sequence of genes, and intragenetic regions (introns) |
| RNA | Gene is turned on or off, variations in RNA sequence |
| Protein | Structure, function |
| Methylation, histone, chromatin | Gene silencing, epigenetic changes, gene packaging |
| Metabolites | Carbohydrates, lipids, and other metabolites that change in response to cell processes |

Human DNA can be isolated from a variety of specimens, as long as there is DNA in the cells (red blood cells do not have DNA, but most human cells do). Depending on the quality of DNA required and the test, very small amounts of blood – such as spots on filter paper, saliva, or buccal cells can be sufficient for detecting human gene variants, which can then be related to disease and the presence of specific microbes or groups of microbes.

Microbial DNA can be detected in all types of samples, making the question of whether a particular microbe was present, although not whether it was alive, fairly easy to answer. Obtaining enough DNA for more extensive testing requires growing the microbe in the laboratory. There are many different microbes, each of which has slightly different growth requirements in a laboratory setting. Further, the number of microbes present can vary tremendously by microbe and among specimens. Protozoa occur at much lower density than bacteria, but are much larger. Thus, the ability to identify microbes by genetic sequence alone, without growing it first, has enhanced detection in clinical and epidemiologic applications. Collecting and storing specimens for detection of microbes rather than detection of viable microbes by growing them in the laboratory greatly simplifies field operations. Specimens for many tests can be dried onto filter paper, and stored at room temperature. Molecular techniques also increase the value of stored specimens, because the microbe of interest still can be detected even if it died during storage. Further, the presence of multiple microbes and their relative abundance can be detected giving a snapshot of the microbiota in a specific site. Moreover, studies can detect the presence and relative abundance of microbial virulence or resistance genes or genes for specific processes independent of the species present, enabling testing of hypotheses regarding the role of specific microbial functions on disease. This capability opens the door for answering questions regarding potential cooperation and antagonism among microbes and between microbes and their host.

Messenger RNA is present when a gene is being transcribed, that is, when protein is being made. This is called "gene expression," and mRNA is sometimes called "transcript." The detection of mRNA enables the determination of which genes are activated in response to infection and gives clues as to gene function. Gene expression is regulated; it can always be on, called constitutively expressed, or it can be temporarily or permanently silenced. Changes in gene regulation can occur in response to environmental cues; bacteria may switch on or off virulence factors. Environmental exposures, including infection, can change the expression of human genes, changing risk of disease. Monitoring which genes are expressed as a pathogen is transmitted, colonizes, and causes disease can be extremely useful in identifying mechanisms to prevent or minimize these processes. Similarly, monitoring the human host response to pathogen acquisition, growth, and invasion can help identify novel therapeutics.

The physical and biochemical characteristics that result once protein is produced is called the phenotype. A microbe might be detected by a genetic sequence unique to its species, the genotype, or by its morphology and response to various biochemical tests, the phenotype. Protein structure largely determines protein function, so studies of structure are essential for understanding the underlying biology and for identifying targets for intervention. At present, our ability to predict protein structure from genotype is minimal, and the correlation between the amount of mRNA expressed and protein produced is poor. Measuring protein directly is desirable, although much more challenging. Proteins indicate host response to infection with a specific microorganism, and vice versa. Once identified, proteins corresponding to different disease stages can be used for diagnosis and as prognostic factors.

Many small molecules (metabolites) such as lipids and carbohydrates vary in health and disease. Several chronic diseases, such as diabetes, cardiovascular disease, and bone tissue loss, are monitored using biomarkers of metabolism. In the future, it is likely that metabolites will give additional information on the host–microorganism interaction that will be useful diagnostic tools and prognostic indicators.

## 5.3 GENE VARIANTS, SNPS, INSERTIONS, DELETIONS, AND FRAMESHIFT MUTATIONS

Epidemiologists look for determinants of different population distributions of disease. One determinant is variation in human genes that are associated with ability to resist infection or fight off infection once it occurs; a second is variation in the genes of the microbes causing infection that make them more or less virulent. Both human and microbial determinants of disease distributions can be detected by looking for gene variants. Among humans and other eukaryotes (including fungi) and in many viruses, genes are in the same physical locations. Variations in the genetic code of a specific gene that occur in the same physical location are called alleles. In bacteria, variations in gene sequence are called variants, as there is not always a consistent physical location for a gene.

During cell replication, minor changes in the genetic sequence can occur that result in mutations. Mutations that do not interfere with function may not be repaired, for example, if they occur in a noncoding region. Spontaneous mutations can occur at a predictable rate, which can be used as a molecular clock for gauging the time frame of evolutionary history. Some regions within the genome of some organisms are more prone to variation, and some organisms have faster mutation rates than others; RNA viruses are particularly prone to transcription errors.

Many genes have known variants that result in slight differences in the resulting protein. Some *Escherichia coli*, bacteria found in the gastrointestinal tract that can also cause a variety of diseases, have surface proteins known as adhesins that enable them to attach to tissues. Gene variants coding for one such adhesin, *Dr*, allow the *E. coli* to bind to slightly different parts of the same molecule. In humans, the secretion of blood group antigens into saliva is controlled by the *Se* gene.[2] Humans who have a silent *Se* gene do not secrete blood group antigens into saliva or in other tissues; having a silent *Se* gene is associated with increased risk of a variety of infections, including *Helicobacter pylori*,[3] which causes peptic ulcer.

### 5.3.1 Single Nucleotide Polymorphisms

The alphabet for amino acids consists of triplets of nucleotides. There are more possible combinations than amino acids, making it possible for more than one code to exist for a given amino acid, and for many amino acids this is the case. For this reason, it is said that the genetic code is redundant. Generally the redundancy is in the last nucleotide in the triplet. For example, regardless of the nucleotide in the third position, if the first two nucleotides are CC the sequence codes for the amino acid proline (CC**U**, CCC, CC**A**, or CC**G**). For proline, if there is a change in the third position there is no change in the

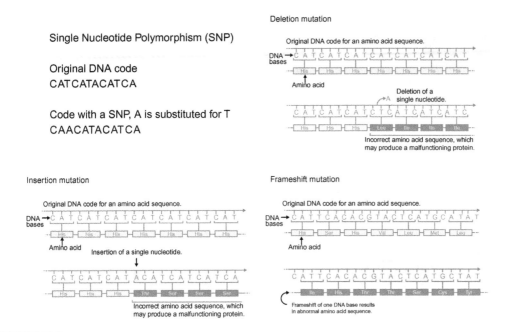

**FIGURE 5.2**
Examples of single nucleotide polymorphisms (SNPs), insertion, deletion, and frameshift mutations. *Source: Reproduced from the U.S. National Library of Medicine.*[4–6]

amino acid; this type of change is called synonymous. However, changing the second nucleotide in the codon for proline changes the amino acid: CAU and CAC code for histidine, whereas CAA and CAG code for glutamine. A change in sequence that results in a change in amino acid is called nonsynonymous. Changes in a single nucleotide are called single nucleotide polymorphisms or SNPs, pronounced "snips" (Figure 5.2). Polymorphism comes from two Greek words, *polloi*, which means "many," and *morphe*, which means "form." Polymorphism therefore means "many forms." A SNP can result in a synonymous or nonsynonymous change in sequence. SNPs can occur in coding and noncoding regions of the chromosome. Synonymous SNPs and SNPs in noncoding regions are useful for distinguishing between microbial strains and determining genetic origin (phylogenetic analysis).

## 5.3.2 Insertions, Deletions, and Frameshift Mutations

In addition to mutations that result in a change in a single nucleotide, one or more nucleotides might be inserted or deleted (Figure 5.2). Insertions and deletions are called "indels." An indel in a noncoding region can be informative for typing and phylogeny. Inserting or deleting a nucleotide in a coding region can shift the base where the code is read. Because the code for an amino acid is a triplet, beginning to read the genetic code at base 2 rather than base 1 results in an entirely different protein. When this occurs, the resulting mutation is called a "frameshift" mutation.

## 5.3.3 Other Types of Gene Variation

Other types of gene variations also are used to distinguish between microbial strains, in phylogenetic analyses, and to detect human and microbial genetic variants. These include: regions of DNA where there are short repeated sequences of nucleic acids (microsatellites), variable number of tandem repeats (VNTR), or an insertion or deletion of genetic sequence that changes the observed patterns in the size of gene fragments when cut with a restriction enzyme and run on a gel (restriction fragment length polymorphisms or RFLPs).

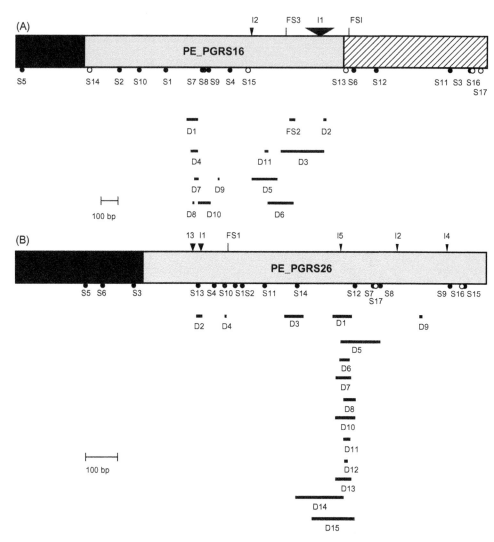

**FIGURE 5.3**

Map of the positions of different sequence variations found in the *M. tuberculosis* PE_PGRS16 (Panel A) and PE_PGRS26 (Panel B) genes. The PE domain is shaded in black and the PGRS domain is shaded in light gray. The striped region in Panel (A) represents an additional atypical sequence at the carboxy-terminus of PE_PGRS16. Triangles represent insertions, bold lines represent deletions, open circles represent synonymous SNPs, and solid circles represent nonsynonymous SNPs. The asterisk in Panel (A) labeled FS2 represents a 1-bp deletion. *Source: Reproduced, with permission, from Talarico et al.[7] (2008).*

While the genome of *Mycobacterium tuberculosis*, the microbe that causes tuberculosis, is generally fairly stable, it has some regions of high variability.[7] These variations may be associated with the ability of *M. tuberculosis* to persist within the host. In a study of the highly variable PE_GRS region of the *M. tuberculosis* genome, Talarico and colleagues[7] observed several different mutations (Figure 5.3) among 200 *M. tuberculosis* strains. Half of the strains had in-frame insertions and deletions, frameshifts, and SNPs sequence variations in two genes found in this region, *PE_GRS16* and *PE_GRS26*.

## 5.4 EXTRACHROMOSOMAL ELEMENTS AND TRANSPOSONS

Human cells differ from bacterial cells in a variety of ways. Bacterial cells usually have cell walls, while human cells do not. Most bacteria have a single circular chromosome; human cells have pairs of chromosomes that are stored in a complex of linear DNA and protein, called chromatin, which must be unwound to be transcribed. Bacterial cells do not have a nucleus, nor do they have organelles, "little organs" with specific functions. A very important organelle found in all human and all eukaryotic cells is the mitochondrion. Lastly, bacterial cells can have extrachromosomal elements, which are not found in human cells.

A type of extrachromosomal element found in most bacterial species is called a plasmid. Plasmids are extrachromosomal elements that self-replicate. Plasmids can carry a variety of genes, notably, genes that code for antibiotic resistance. They also can be transferred between bacteria via horizontal gene transfer (described in Section 5.6). A pathogenicity island is a region of the chromosome coding for virulence genes whose sequence suggests it was acquired by horizontal gene transfer; it may have been a plasmid that incorporated into the chromosome. The genes coding for Shiga toxin in the food-borne pathogen *E. coli* O157:H7 are in a pathogenicity island. Pathogenicity islands are detected by comparing the percent of guanine–cytosine (GC) found in the island to that found in the rest of the chromosome; GC content tends to vary by species. Once acquired, a pathogenicity island is copied during replication with the rest of the genome.

A transposon is a piece of DNA that has recognizable genes that inserts itself into the chromosome or a plasmid. Transposons tend to move around the genome, but also can be passed between bacteria. Because they move around, they are sometimes called "jumping genes." The number and location of transposons can vary across the bacterial genome; this makes them useful for distinguishing between bacteria of the same species. The IS6110 transposon is the basis of one of the typing systems used for *M. tuberculosis*. A special type of transposon has a gene capture system; these are known as integrons. Integrons are found on bacterial chromosomes, on plasmids, and transposons. The gene capture system in integrons enables it to take up DNA free in the environment in a form known as a gene cassette, and integrate it into the genome and express the gene. The most common gene cassettes identified to date are for antibiotic resistance.[8]

## 5.5 RECOMBINATION

Genetic reassortment, the mixing of genes between two organisms to make a new genetic sequence known as a recombinant, is a powerful mechanism for evolution and adaptation. Sexual reproduction genetically recombines the genes of each parent. Each human is a recombinant of the parents' genes. Fungi reproduce sexually and asexually; when fungi reproduce sexually, the offspring are recombinants. For organisms that do not reproduce sexually, recombination also occurs, but is independent of reproduction.

Sexual reproduction results in large scale rearrangement of DNA during the process of meiosis. Diploid organisms have two copies of each chromosome, with each chromosome carrying a complete set of the same genes. These genes occur in the same order and in the same physical location. Meiosis results in a reassortment of the maternal and paternal chromosomes; the resulting chromosomes have a mix of genes from each parent. Figure 5.4 shows this process for humans; each of the two chromosomes in the daughter cells (meiosis II) are a mix of the original black and white chromosomes, recombinants.

Viruses do not have a cell and are dependent on the cells they invade to reproduce. However, viruses can create recombinants via a process called genetic reassortment. If two of the same type of virus enter a single cell the genes may genetically reassort, resulting in recombinants. Viruses that infect bacteria, known as bacteriophage or phage, can also pick up bacterial DNA during lysis. Genetic reassortment causes major changes in the influenza A virus, resulting in a genetic shift in the virus and epidemics. Recombination occurs among bacteria by a process known as horizontal gene transfer (described in Section 5.6), which also has been observed among virus and fungi. The creation of bacterial recombinants is an important tool in the molecular biologists toolbox.

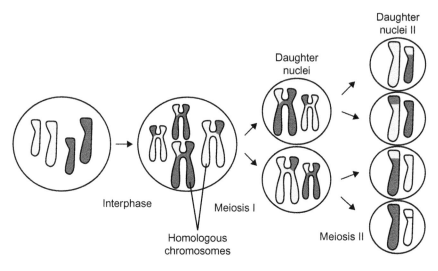

**FIGURE 5.4**

Meiosis, a type of nuclear division, occurs only in reproductive cells and results in a diploid cell (one having two sets of chromosomes) giving rise to four haploid cells (having a single set of chromosomes). Each haploid cell can subsequently fuse with a gamete of the opposite sex during sexual reproduction. In this illustration, two pairs of homologous chromosomes enter *Meiosis I*, which results initially in two daughter nuclei, each with two copies of each chromosome. These two cells then enter *Meiosis II*, producing four daughter nuclei, each with a single copy of each chromosome. *Source: Reproduced from A Basic Introduction to the Science Underlying NCBI Resources.*[9]

## 5.6 HORIZONTAL GENE TRANSFER

Genetic mutations can occur in response to an environmental exposure, or during reproduction. In microbes, there is an additional mechanism for genetic mutation: the acquisition of DNA in a process called horizontal gene transfer. Horizontal gene transfer (also known as lateral gene transfer) is the introduction of genetic material into another organism that is not its offspring. The material can be a plasmid, a transposon, or other genetic material. Horizontal gene transfer has been detected in viruses, bacteria, and fungi, and the transfer is not limited to within a species or even within a kingdom: β-lactam synthesis genes (genes that make bacteria resistant to β-lactam antibiotics such as penicillin) are hypothesized to have been transferred from bacteria to fungi.[10] The ability to artificially introduce genetic material from one organism into another has three important applications: gene cloning (obtaining multiple copies of genetic material for further study), genetic manipulation, and the study of the function of particular genes. The artificial system where most experimental horizontal gene transfer has been conducted is in bacteria, most frequently in *E. coli*.

Horizontal gene transfer occurs by three different mechanisms: transformation, transduction, and conjugation. This is shown in Figure 5.5 for a gene coding for antibiotic resistance *(Ab$^r$)*. Naked DNA can be taken into the cell directly from the environment and be incorporated into the bacterial genome in a process known as transformation. Some bacteria have a tendency to transform, particularly when stressed, for example, when exposed to antibiotics. If a bacterial cell is naturally transformative, it is called "competent." Competent bacterial strains are extremely useful for laboratory experiments. Competency is also a powerful evolutionary strategy for responding to a changing environment. More than 60 species of bacteria are naturally transformative, including several major human pathogens.[11]

Horizontal gene transfer of bacterial DNA into a bacterial cell by a virus is called transduction. Viruses that infect bacteria are called bacteriophage or phage. When a virus infects a cell, it takes over the cell machinery to replicate itself. Sometimes during the packing of viral DNA into the phage head there is an error, and bacterial chromosomal

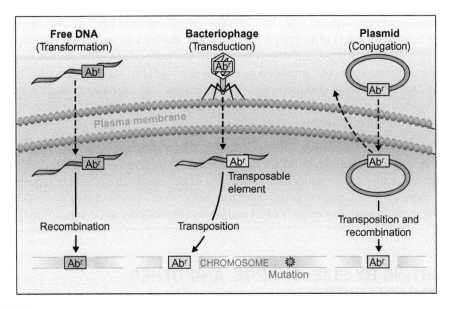

**FIGURE 5.5**
(1) Transformation: uptake of free DNA into the cell; (2) Transduction: phage introduces DNA into the cell. (3) Conjugation: transfer of plasmids between cells. *Source: Reproduced, with permission, from Alekshun and Levy[12] (2007).*

material is included. Other times the virus has been inserted into the bacterial chromosome. When it removes itself, sometimes it includes pieces of the bacterial chromosome. In either case, when the virus infects a new bacteria cell, the bacterial DNA may be integrated into the bacterial chromosome, or if it was a plasmid, it remains as extrachromosomal material. Extrachromosomal material also can be transcribed with resulting phenotypic changes. Transduction is used to artificially introduce genetic material into mammals and plants; this process is called transfection.

Genetic material is also exchanged directly between bacteria in a process called conjugation. Bacterial mating may occur via a sex pilus or other structure. The stimulus for exchange can be from the plasmids or transposons or an integrative conjugative element (ICE). ICEs are pieces of DNA with known genes that insert themselves into the chromosome, have the machinery to remove themselves from the chromosome, replicate, and transfer a copy to another bacterial cell. Plasmids can carry genes for antibiotic resistance and other virulence factors; conjugation is thought to be an important mechanism for exchange of antibiotic resistance genes among bacteria.

Horizontal gene transfer enables bacteria to acquire very large pieces of DNA. This makes it possible for the bacterial genomes of members of the same species to vary dramatically in size: the genome of *E. coli* can be as small as 4 million base pairs and as large as 6 million base pairs. *E. coli* O157:H7 is a food-borne pathogen that causes bloody diarrhea and hemolytic uremic syndrome. The *E. coli* O157:H7 toxin is very similar to that found in *Shigella*; the genes for Shiga toxin are carried by phage.[13] Analyses suggest that *E. coli* O157:H7 acquired its toxin-making abilities following infection by a phage.

## 5.7 AN INTRODUCTION TO COMMON MOLECULAR METHODS

Molecular tools commonly used in molecular epidemiologic studies of infectious diseases can be roughly classified into three groups: (1) methods based on sorting by size, charge, or other characteristics (gel electrophoresis, flow cytometry, mass spectrometry); (2) methods based on the polymerase chain reaction (PCR), which enables the amplification of DNA or RNA and facilitates manipulation and sequencing; and, (3) methods using hybridization/antigen–antibody reactions. This section presents a brief introduction to these techniques followed by an example of a specific application in an epidemiologic study and a discussion of the advantages and disadvantages of the particular technique for population studies. The presentation is not intended to enable the reader to actually conduct the techniques, but to understand something of the strengths and limitations of their use in an epidemiologic context, and thus serve as a springboard for discussion when planning and conducting a molecular epidemiologic study.

The intent is not to be exhaustive – the technology is in a constant state of development – but to provide sufficient background for the naïve reader or a review for the more knowledgeable reader of the terms used in the rest of the book. The reader is referred to the many more complete presentations available both on the Internet (see, for example http://www.molecularstation.com[14]) or in textbooks (*From Genes to Genomes*,[15] for example).

## 5.8 SORTING BY SIZE, CHARGE, AND OTHER CHARACTERISTICS

Classification is the first step toward scientific discovery. Tools of molecular biology bring classification to a new level. Whereas in the past microorganisms would be classified by shape, size, and the results of biochemical assays, molecular tools classify based on DNA or RNA sequence; the size, charge, structure, or other characteristics of proteins or metabolites; and surface molecules expressed by cells. Sorting organisms into groups based on these characteristics enables description and association of these substances with health and disease (Table 5.2). Sorting methods can be used to classify infectious agents into more specific groups. This increased discrimination enables the investigator to determine if cases with a specific infectious disease transmitted to each other or acquired infection from a common source, or to differentiate epidemic from endemic strains. The patterns resulting from sorting rRNA of bacteria present in the mouth might be compared over time to comment on the stability of oral microbiota. Or sorting may be a first step toward further detection: the presence of a protein of a specific size might occur more often among those with than without disease. Detection of this protein might lead to a diagnostic or prognostic test or increased understanding of the disease mechanism.

**TABLE 5.2** Sorting Methods, Material Sorted, and How Sorted

| Method | Material Sorted | Sorted By |
| --- | --- | --- |
| Electrophoresis | DNA<br>RNA<br>Protein | Size |
| Gradient gel electrophoresis | DNA | Size<br>Sequence |
| Sodium dodecyl sulfate Polyacrylamide gel electrophoresis (SDS-PAGE) | Protein | Size (length and molecular weight)<br>Structure |
| Flow Cytometry | Cells | Size<br>Phenotype |
| Microfluidics | Cells<br>DNA<br>RNA | Size<br>Sequence<br>Phenotype |
| Matrix-assisted laser desorption ionization (MALDI) (a type of mass spectrometry) | Protein<br>Metabolite | Mass |

## 5.8.1 Electrophoresis

Electrophoresis is a well-developed technique to sort DNA, RNA, proteins, and carbohydrates. The sorted material can be visualized and further identified, if required. Electrophoresis can be used to visualize the PCR product resulting from amplifying a specific gene or a purified protein. In this case, the investigator is looking for a band of a specific size, often corresponding to disease or exposure presence. Alternatively, the investigator might be looking for the pattern that results from cutting purified DNA from a single microbe, or from PCR products from amplifying 16S rRNA from a community of microbes, providing a unique "fingerprint" of a specific microbe or microbial community (Table 5.3). Band patterns resulting from cut DNA can be analyzed using phylogenetic techniques to determine evolutionary relationships.

Products can be sorted by size, conformation (three-dimensional form), and guanine (G) and cytosine (C) content. Electrophoresis refers to the use of an electric current to separate particles by charge (Figure 5.6). A typical application is separating DNA that has been cut into pieces. Determining size requires a way to sort the pieces and a measuring stick. Pieces of DNA are placed in a hole known as a "well" at the top of a slab of gelatin of known density (called a "gel"). Because DNA has a negative charge, if a current is run through the gel, the DNA moves toward the positive end (bottom of the slab). The pieces move through the pores in the gel at a rate proportional to size; thus larger pieces move more slowly than smaller pieces. For very long pieces, the current may be applied in a pattern that alternates the direction ("pulsed") helping longer pieces to snake through the gel. After a gel is "run," the smallest pieces are at the bottom of a "lane" and the largest pieces are at the top. The size is determined by comparing to nucleic acid markers of known size, which are run in one or more lanes on the gel. Known as a "ladder," the markers are used as a ruler to determine the size of the substance of interest. The gel is stained to visualize the bands, or the DNA

63

---

**TABLE 5.3 Applications of Electrophoresis to Molecular Epidemiology**

- Diagnosis/species identification
- Molecular fingerprinting/microbial community fingerprinting

---

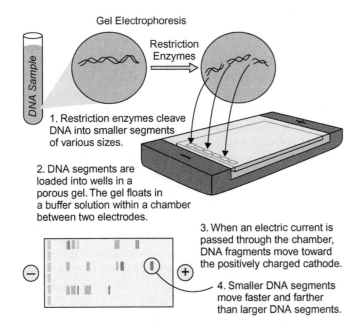

Gel Electrophoresis

Restriction Enzymes

DNA Sample

1. Restriction enzymes cleave DNA into smaller segments of various sizes.

2. DNA segments are loaded into wells in a porous gel. The gel floats in a buffer solution within a chamber between two electrodes.

3. When an electric current is passed through the chamber, DNA fragments move toward the positively charged cathode.

4. Smaller DNA segments move faster and farther than larger DNA segments.

**FIGURE 5.6**

Electrophoresis. *Source: Reproduced, with permission, from Huntington's Outreach Project for Education at Stanford.*[16]

is transferred to a membrane and hybridization techniques (described below) are used to detect bands that contain a sequence of interest. Bands of interest may be literally cut out of the gel and sequenced or tested in other ways.

RNA, proteins, and carbohydrates also may be separated based on charge or conformation using electrophoresis. Because of their larger size and more complex structures, proteins often are separated based on conformation using a two-dimensional gel. A two-dimensional gel results in a matrix, with each substance categorized by two different separation reactions. A mixture of proteins of various sizes might be placed in the well at the top of a single lane. Current is first run so the proteins are separated from top to bottom. The proteins are now spread out in a line. To further resolve the groups, the current is run from side to side.

Gels may also include a temperature or chemical gradient, so that the temperature or chemical concentration increases from the top to the bottom of the gel. The temperature or chemical gradient causes double-stranded DNA to become single-stranded (denature) as it moves through the gel with a gradient. Denaturing causes DNA, RNA, and protein to lose structure. Structure modifies how a material moves through a gel – imagine pulling a ball through water rather than the ball cut into one long strip. A tightly coiled structure moves much faster than if it is unwinding. If the DNA to be sorted is double-stranded, denaturing causes it to separate into two strands, which slows its progress through the gel. This makes it possible to sort double-stranded DNA by sequence, because sequence with more bonds will take longer to denature. Recall that nucleotides guanine and cytosine bind together with 3 hydrogen bonds whereas adenine and thymine have 2 hydrogen bonds. Thus DNA with higher guanine–cytosine content denatures more slowly. Pieces that denature fastest move the slowest. As the strands separate they fan outward, increasing drag.

Gels can be made as a slab or put in a capillary tube. Capillary tubes provide very high resolution, so that proteins that differ by one amino acid or DNA that differs by one nucleotide can be separated. The resolution is higher within a capillary tube because higher voltages can be used. However, slab gels are more useful for identifying patterns. Capillary tubes are used in automated applications, such as sequencing.

The patterns observed after cutting bacterial DNA with enzymes at specific sites (restriction sites) are the basis for using pulsed-field gel electrophoresis (PFGE) as a bacterial typing technique. PFGE is currently used by the Centers for Disease Control and Prevention in its surveillance of food-borne bacterial infections. The ability of PFGE to distinguish between isolates depends on the number of bands obtained; and different enzymes will give different patterns when applied to the same bacterial isolates. This can cause problems in interpretation if two isolates have the same pattern using one enzyme but different patterns using another. Results from PFGE are also used in phylogenetic analyses. Other bacterial techniques also are based on visual comparison of the patterns observed after running PCR products or PCR products cut with a specific enzyme on a slab gel. Diagnostic kits based on PCR generally use electrophoresis to detect the presence of a PCR product of a specific size that indicates the substance has been detected (PCR is described in Section 5.9).

Techniques that depend on visual patterns have a number of drawbacks. Slab gels can break or crack. The content and type of gel makes a difference in the quality of resolution, and must be tailored to the material to be resolved. To run current through a gel, it must be placed in a buffering solution, which must be properly prepared. If the current is too high or run too long the gel may melt. Running the current for too long can run the material off the gel entirely. There is a limit to how big a slab gel can be, limiting the number of samples that can be screened at a time. The bigger the gel, the more markers that need to be included, as position on the gel can influence the speed a substance travels, for example, if the gel is not of uniform thickness. The quality of visual patterns that appear on a gel vary. Sometimes bands are strong, other times they are faint. There can be smears. Gel-to-gel variation limits the validity of across-gel comparisons. Although there are computer programs that normalize band patterns to that of a standard (markers that indicate band size and/or a reference isolate), the user has considerable control over how this normalization takes place. Bands may be included or excluded, which may substantially change results. Thus it is best to include all samples of interest on the same gel.

The strengths and limitations of electrophoresis in an epidemiologic context depend on whether electrophoresis gives the end product, or a step toward an end product. The result of a test might be that there is a band of a specific size, or two bands for a positive test, one band for a negative test. In this case, the results are (relatively) easy to interpret, particularly if all proper controls are included. As noted earlier, results based on visual patterns are somewhat more problematic, especially if large numbers of specimens are to be tested. As a technique for separating substances for further testing, electrophoresis is a robust method that, as long as proper techniques are followed, can be used successfully in clinical and epidemiologic settings.

## 5.8.2 Flow Cytometry and Microfluidics

Flow cytometry enables the analysis and sorting of cells by size, complexity, and phenotype. This is an extremely powerful technology with numerous applications. One way to think of flow cytometry is as a microscope directed at a fast moving, single file line of cells. Each cell is examined as it flows by, so the relative size can be assessed, and other distinguishing characteristics noted, such as granularity. If some aspect of the cell is marked, with a stain or an antibody, this can be determined also. As the cells pass by, the numbers with each characteristic can be counted, so the proportion of the sample with cells of each type can be assessed. If a cell is of interest, it might be directed to go to a different line; in this way, cells can be sorted into groups for further testing. This assessment requires a flow cytometer, which has three parts: fluidics, optics, and electronics. The fluidics puts the population of particles or cells in the sample into the single file line of "flow." The optics are a set of lasers directed at the cells, and filters that direct the scattered light to detection systems. The electronics converts the light signals into signals that can be read by a computer.

Flow cytometry has long been used in clinical laboratories for determining total lymphocytes, monocytes, neutrophils, and granulocytes in a human blood sample, important diagnostic and prognostic indicators. For infectious diseases, flow cytometry has been used to rapidly detect and collect cryptosporidia oocysts for further typing during an outbreak investigation,[17] to characterize immune function in response to disease,[18] and to enhance diagnosis.[19] The key disadvantage of flow cytometry is the cost of the equipment and of equipment use, which has limited its broader application.

Cell sorting and analysis can also be conducted in microfluidic systems. Microfluidic systems deal with extremely small amounts of fluids; DNA chips and laboratories on a chip are examples of microfluidic systems. Many of the processes conducted by a flow cytometer can be done using microfluidics. Microfluidics systems can also sort DNA, RNA, and protein in ways similar to electrophoresis, and microfluidics gene sequencing is available. As laboratory on a chip systems fall in price, there are numerous potential applications for epidemiologic field studies. The systems are developed for use by individuals with minimal training, and analyses can be conducted rapidly, on-site in resource-poor settings. Systems can analyze a drop of blood for the presence of bacteria, fungi, and parasites; the system can identify the microbe and even determine the strain type.[20] This capability greatly enhances field studies. Finger sticks are generally more acceptable than drawing blood; there is no need for storage and transport of the blood, and study participants can be given their results immediately, enhancing participation.

### 5.8.3 Matrix-Assisted Laser Desorption Ionization

Matrix-assisted laser desorption ionization (MALDI) is used to identify proteins. The sample is vaporized and ionized; the time it takes for the ions to travel to the detector is measured to determine the mass of the protein. The process is called matrix-assisted, because the sample is mixed with a matrix, which transfers the laser energy to the sample.[21] In molecular epidemiologic studies, MALDI has been used to characterize plasma products of individuals with and without delirium,[22] identify protein changes associated with human hepatocellular carcinoma due to hepatitis as opposed to nonhepatitis,[23] and to classify microbes into genotypes.[24]

## 5.9 POLYMERASE CHAIN REACTION

In a nutshell, PCR amplifies DNA, and can be used to amplify specific genes (Figure 5.7). PCR is exquisitely sensitive in a laboratory sense, that is, PCR can detect extremely small

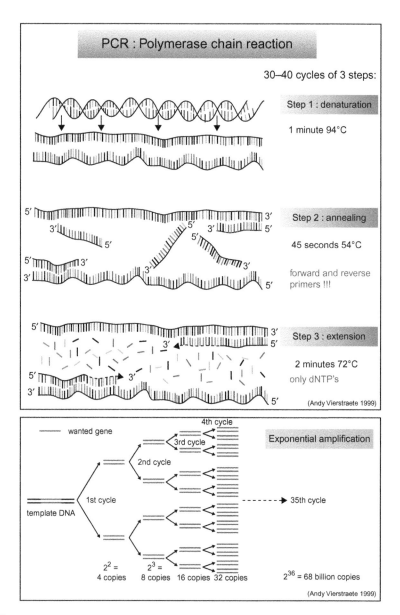

**FIGURE 5.7**

Polymerase chain reaction (PCR). *Source: Reproduced, with permission, from Veirstraete*[25] *(1999).*

amounts of genetic material. The applications of PCR are very broad, from diagnosis to determining genetic sequence. Common applications in molecular epidemiology include molecular fingerprinting, diagnosis, species identification, genetic sequencing, and comparison of gene expression (Table 5.4). In a process called reverse transcription PCR, complementary DNA (cDNA) is first created from RNA, then the cDNA is amplified using PCR, enabling detection of gene expression. PCR products can be visualized using electrophoresis. The sequence of a PCR product can also be determined by detecting the energy released during incorporation of nucleotides, as is done in pyrosequencing. The results of PCR are semiquantitative, essentially present or absent; however, real-time (also known as quantitative) PCR enables relative quantification of the amount of product in the original sample by measuring the amount of amplified product after each step.

PCR begins by denaturing double-stranded DNA, usually by heating, so it separates into two single strands. This is also called "melting." To begin amplification, short segments of DNA sequences, called primers, which genetically complement the beginning of the template sequence are needed. Recall that the sequence of each strand is different, because it contains nucleic acids that complement the other strand; therefore, a separate primer is necessary for each of the complementary strands. Primers are annealed to the single stranded DNA, usually by lowering the temperature. DNA polymerase binds to the primer-template and begins DNA synthesis. DNA polymerase is an enzyme that helps replicate DNA. DNA synthesis is allowed to proceed for a short time before the mixture is heated again denaturing the newly formed double stranded DNA, so that twice as many copies can be made during the next cycle. The cycle is repeated, until the desired amount of amplified DNA is present. Because the amount of DNA doubles each cycle, 25 PCR cycles results in 16,777,216 copies of a single genetic sequence.

## 5.9.1 PCR Applications

PCR amplification is just the first step for many applications. For example, the PCR product may be sequenced, sized, or further processed to detect a disease marker or virulence gene or identify a microbe. The presence of a PCR product of a specific size can correspond to the presence of a biomarker of disease. Some tests require the PCR product to be cut with an enzyme that cuts only when it sees a specific genetic sequence known as a restriction site. When run on a gel, the presence of two bands indicates the biomarker of interest is present. The PCR product, with a label attached, may be used as a probe for dot blot hybridization (see later). In another variant, a test is based on amplifying a probe that complements the sequence of interest. Called ligase chain reaction, two probes are used, each probe binding to a complementary sequence. If the two exactly match the template sequence, they are ligated. Amplification occurs only on the ligated probe, making this a very sensitive and specific test. PCR is also the basis for several microbial fingerprinting methods.

The ability to amplify very small amounts of genetic material means that PCR can be used to detect microbes in a specimen or human DNA in tissue and even microbes or human DNA in the environment. PCR can be used to detect material that is from organisms that are dead, such as DNA from organisms dried on a slide and organisms that are difficult

**TABLE 5.4  Applications of the Polymerase Chain Reaction (PCR) in Molecular Epidemiology**

- Molecular fingerprinting/microbial community fingerprinting
- Diagnosis
- Species identification
- Determining genetic sequence
- Differential display (comparison of gene expression from different groups)

to grow, and microbes that cannot live with oxygen or are obligate intracellular parasites (viruses). Further, PCR can detect genetic material from organisms we don't even know exist, and amplify genetic material from multiple organisms in the same sample. This is a huge advance. Before PCR, either sufficient sample had to be taken for the desired test, or if it was a microorganism, the organism had to be grown. Dead microorganisms were difficult if not impossible to accurately identify. To grow a microorganism requires knowledge of the required growth conditions; our knowledge of growth conditions remains limited to only a small segment of the estimated number of microbial species. Previously, if an organism could not be grown directly, it was necessary to clone the DNA in order to get enough DNA to study. Cloned in this sense means inserting the genetic material into an organism, often *E. coli*, that can make multiple identical copies of the genetic material. Cloning can be a lengthy and time-consuming process.

Microbial fingerprinting methods are used in outbreak investigations as part of a case definition, or as a way to confirm transmission or the infection reservoir. PCR based methods for microbial fingerprinting fall into two general categories. The first is looking for repeats in genetic sequence. These can either be VNTRs, or longer sequences found throughout the genome. One method, random amplification of polymorphic DNA (RAPD), amplifies random sequences. Ideally, these methods give multiple PCR products of different sizes that form a pattern when run on a gel. The band patterns constitute a fingerprint. The other method involves looking for variation in sequence in known regions of the genome. A specific region or set of regions is amplified; they may be either sequenced or cut using a specific enzyme, so that different variants result in cut pieces of different sizes.

Rather than fingerprint a specific organism, PCR can be used to create a snapshot of the entire microbial community, called a community fingerprint. Community fingerprints generally amplify a sequence that is found in all organisms of interest. For example, all cells have ribosomes, but the size and number of copies of DNA sequence coding for the ribosomal RNA (rRNA) varies. Sequencing the rRNA can identify most bacteria to the genus level. Using universal primers for the rRNA (16S for bacteria) to amplify a specimen from a community gives a fingerprint reflecting the frequency and relative abundance of the diversity of bacteria present. Fingerprints can be compared over time and space to determine the bacteria that are most stable or most variable. Although not perfect, using PCR has tremendous advantages over culture techniques for describing microbial communities. First, even if limited to bacteria, there is a great range in growth requirements from differences in nutrition to temperature and oxygen level, making direct culture extremely time-consuming, and still likely to miss organisms. Second, even if growth requirements are similar, microbes grow at different rates, so those that grow faster will outcompete the others. Thus it is difficult to determine relative abundance and frequency. Third, some organisms need others to grow and will therefore be missed if attempts are made to grow in pure culture.

Diagnosis and species identification are the most common clinical uses of PCR. PCR diagnostic tests are more sensitive than other methods because very small amounts of material can be detected. The test can be made more specific by doing PCR in two steps; the first is amplifying a genetic sequence that should contain the sequence on interest. A second step amplifies only the specific sequence of interest. Because there is more template to begin with, the possibility of detecting the sequence, if it is there, is greater, while decreasing the chances of a false positive. PCR methods for diagnosis of infectious agents generally rely on detecting the presence of genetic sequence unique to the organism. Some viruses are so small that the entire genome easily can be sequenced. Automated sequencing is advancing at a rapid pace, so the length of sequence that can be done easily is constantly increasing. For bacteria, identification is generally based on the sequence of genes that code for highly conserved genes, which are species specific.

Because PCR can be made very specific with proper primer design, multiple PCR reactions can be carried out in the same test tube. This is known as multiplex PCR. For example, a multiplex PCR has been developed that, in a single experiment, detects *Staphylococcus*, distinguishes between *Staphylococcus aureus* and coagulase-negative *Staphylococcus*, and determines whether the *Staphylococcus* has genes that code for oxacillin resistance.[26] Another multiplex PCR screens samples simultaneously for all four species of *Plasmodium* that cause malaria, rather than in separate reactions.[27] Development of multiplex PCR can be challenging, as the primers must be developed in such a way that they neither interfere nor react with each other. Further, it can be time-consuming to determine the optimal conditions, as each individual reaction may require slightly different conditions. Thus the time to develop a multiplex PCR must be weighed against the ultimate savings of doing one versus multiple PCR reactions.

The last application is to compare gene expression between exposed and unexposed or diseased and nondiseased individuals or between microbes in different environments. Comparison of expression signatures among humans can identify diagnostic or prognostic indicators and increase our understanding of disease pathogenesis. A gene expression study comparing prospectively collected specimens from patients with classical dengue to those who progressed to dengue hemorrhagic fever identified gene expression signatures that accurately predicted progression to dengue hemorrhagic fever.[28] A study, comparing liver cell activation profiles among individuals with hepatitis C who did not respond to interferon therapy compared to those who did respond, found that the types of liver cell activated differed between the groups.[29] Among microorganisms, gene expression studies can identify genes associated with transmission, pathogenesis, or virulence. In a mouse model, a secreted esterase of group A streptococcus was identified as an important virulence factor in soft tissue infection.[30] Since gene transcripts are mRNA, the RNA first must be reverse transcribed to cDNA for study. Expression studies also use quantitative PCR (see Section 5.9.3), and microarrays (see Section 5.11.1) to detect transcripts.

PCR is also used in sequencing, which is discussed in Section 5.10.

### 5.9.2 Technical Considerations

A major concern with PCR is that if an error is introduced during amplification, the error will be propagated because the prior amplification product is used as template. The error can be due to amplification of a contaminant or the introduction of a mutation during amplification. Because amplification makes it possible for PCR to detect extremely small amounts of DNA, the erroneous introduction of even a small segment of contaminating DNA can result in detectable (false-positive) signal, depending on the number of amplification steps. Thus extreme care must be taken when using PCR to avoid contamination, and positive and negative controls should always be included. To minimize contamination, pipetting is done in a different location from the amplification. To detect contamination, PCR experiments typically include a control containing only water; if a PCR product is observed in the water control, it is an excellent indicator that contamination occurred. DNA polymerases, the enzymes involved in DNA synthesis, have different error rates. Those used in diagnostic and sequencing applications generally are somewhat self-correcting, with errors occurring at a rate of 1 per 10,000 base pairs. However, errors occur. Typical mutations are substituting one nucleotide for another, and the deletion or addition of one or more nucleotide. The fidelity of copies is of particular concern when PCR is used to detect variants in sequence, such as SNPs.[31] Therefore it is also optimal to use the minimal number of cycles of amplification.

The design of a PCR primer is critically important. If the primer is not designed appropriately, it will not amplify the region of interest or may amplify regions not of interest. A primer can also anneal to itself, resulting in PCR product; this is called a "primer-dimer." PCR is also sensitive to experimental conditions. Amplification can occur even if there is not a perfect match between primer and template. By lowering the annealing temperature, the primer will anneal to the template even if there are some mismatches. This can be beneficial in a diagnostic test, for example, if the starting sequence of the template of interest is somewhat variable. Alternatively, nonspecific binding can result in amplifying DNA that is not of interest. By contrast, if the annealing temperature is too high, there may be little or no product produced. Given these limitations, it is critical to determine if the lack of PCR product represents a true negative or an experimental error. If there is a PCR product, it is also essential to determine that it represents the region of interest.

Almost always, PCR is conducted using a machine known as a thermocycler, which varies temperature according to an investigator-set program. Thermocyclers wear out, are sensitive to ambient temperatures, and vary in quality. Especially for experimental assays, the assay may need to be conducted on the same machine to ensure reliability and reproducibility.

Other errors may arise from the template itself or the reagents used in the reaction. If the DNA is of poor quality, it may not amplify properly. Different specimens require different preparations: some specimens contain molecules that inhibit the PCR reaction, for example, hemoglobin or bile salts. These must be removed before amplification.[32] The primers used can be incorrectly synthesized or diluted; if contaminated with nucleases (for example, from skin) they can be degraded. If PCR is used for microbial community fingerprinting, with a universal primer set, differences in annealing or other technical issues can lead to overamplification or underamplification of specific gene sequences, resulting in biased estimates.

Conventional PCR is very good at detecting small quantities of genetic material, but it is only semiquantitative, that is, it cannot accurately detect the amount of genetic material present. Though detecting presence or absence is sufficient for most diagnostic assays, accurately determining quantity is essential for studies of gene expression and describing microbial communities.

### 5.9.3 Quantitative PCR, Also Known as Real-Time PCR

PCR, as described above, enables detection of the presence or absence of genetic material. But how much genetic material is there? Quantitative PCR, also known as real-time PCR, answers this question. (Real-time PCR is abbreviated RT-PCR, which makes it easily confused with reverse transcription PCR, which refers to copying RNA to DNA. Therefore the term quantitative PCR [qPCR] is preferred.) Because amplification is not perfectly exponential, comparing the amount of DNA obtained following amplification is not a good indicator of total amount of DNA that was present in the starting material. Moreover, in regular PCR the amount of product is generally assessed by running the product on the gel and staining; this is rather insensitive. However, early on during amplification, PCR is close to exponential, doubling each cycle, and small differences in starting material can be detected quickly; after four cycles there is more than a thousandfold difference in the number of DNA copies separating initial starting amounts of 10, 20, 30, and 40 copies (Table 5.5).

**TABLE 5.5** Comparison of Number of Copies Obtained at the End of Each PCR Cycle by the Number of Templates in the Sample

| Copy numbers in sample | Cycle | | | |
|---|---|---|---|---|
| | 1 | 2 | 3 | 4 |
| 10 | 100 | 10,000 | 100,000,000 | 10,000,000,000,000,000 |
| 20 | 400 | 160,000 | 25,600,000,000 | 655,360,000,000,000,000,000 |
| 30 | 900 | 810,000 | 656,100,000,000 | 430,467,210,000,000,000,000,000 |
| 40 | 160,000 | 25,600,000,000 | 655,360,000,000,000,000,000 | 4.294967296E+41 |

Measurement of the amount of DNA during the exponential phase requires a different method of detection as the absolute amount of DNA is small. To detect the DNA, qPCR counts the number of copies obtained in each cycle through a fluorescent reporting system that measures the amount of double-stranded DNA present. To quantitate the test sample, the sample is compared to a serially diluted standard control. Figure 5.8A shows the amount of fluorescence in arbitrary units plotted by cycle number. The measure is most accurate when amplification is truly exponential; this point is set as the threshold, the line at ~800 on the y-axis line parallel to the x-axis in the figure. Figure 5.8B shows the log starting quantity for standards (open circles) by threshold cycle. Notice that the log of the starting quantity is a straight line: amplification was exponential. Unknowns are shown as crosses. By reading the x-axis, the starting quantity can be obtained. qPCR allows the user to check of the accuracy of amplification, by comparing the melting temperate of the PCR products. The temperature required to denature double-stranded DNA depends on the number of GC bonds. Recall, GC has three hydrogen bonds and AT, two hydrogen bonds. All the PCR products derived from the same primer should have the same melting temperature, unless there is contamination or some other error. Figure 5.9 shows the melting curves for the

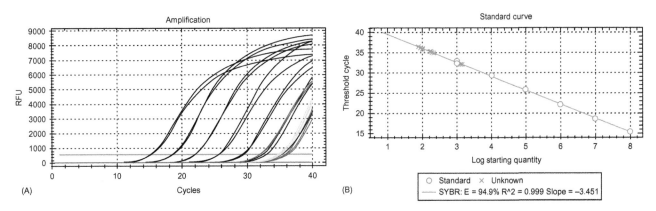

(A)

(B)

**FIGURE 5.8**
Results from qPCR using a universal bacterial primer on bacterial DNA extracted from human milk specimens. **A,** Graph shows serially diluted control and unknowns (line beginning at 30 cycles). Line at ~0 on the y-axis is no DNA control; line at ~800 on the y-axis is the threshold cycle. **B,** Graph shows log of starting DNA quantity by threshold cycle for serially diluted control *(open circles)* and unknowns *(crosses). Source: Author's data.*

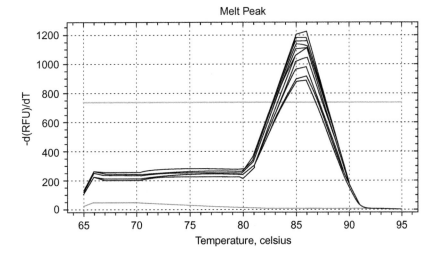

**FIGURE 5.9**
Melting curve from qPCR using a universal bacterial primer on bacterial DNA extracted from human milk specimens.
*Source: Author's data.*

reaction in Figure 5.8. The *y*-axis is in arbitrary fluorescent units, transformed so the melting peak is clear; the *x*-axis is the temperature. The line just above zero on the *y*-axis is a no template control, and the line at ∼800 on the *y*-axis is the threshold. As desired, all the PCR products had the same melting temperature. An informative tutorial on qPCR and analysis is available at http://pathmicro.med.sc.edu/pcr/realtime-home.htm.[33]

## 5.10 SEQUENCING

There are several methods for obtaining the sequence of nucleotides in DNA. The methods have been automated, and new improvements are on the horizon. The length of DNA that can be read varies by method, but no current method has long enough read lengths that the human genome or even a bacterial genome can be read straight through. Thus genome sequencing is done in parallel, with the genome broken into fragments and then the fragments reassembled. When looking for gene variants, only the part of a gene of interest is amplified; this often can be read all at once. DNA (with the exception of single-stranded viral DNA) is double-stranded. Sequencing is usually done on both strands as a validation.

Automated genetic sequencing machines are based on PCR. An example is the 454 Sequencer™ (Roche) , which uses pyrosequencing. The capability of these machines is such that a bacterial genome can be sequenced and the entire sequence assembled within a week. PCR is used to amplify the DNA fragments to be sequenced, to give sufficient DNA for the sequencing reaction. Often the fragments are immobilized on a bead using emulsion (emulsion-based PCR). In pyrosequencing, the chemicals released when a nucleotide is incorporated are converted by enzymes into light, which is then recorded by a camera. The procedure introduces nucleotides one at a time. The amount of light emitted is proportional to the number of nucleotides incorporated (Figure 5.10). In the Sanger method, nucleotides are labeled for detection (a label enables detection usually by giving off light or radiation), but only a small proportion of all nucleotides added are labeled. The labeled nucleotides stop sequence amplification (Figure 5.11). When run on a gel, the shortest fragments correspond to the beginning sequence and longest to the end, so the sequence can be read off the gel.

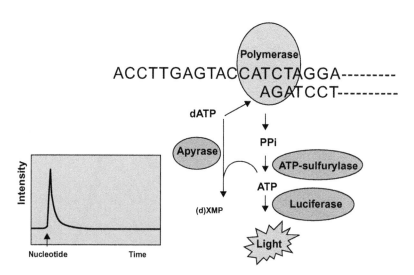

**FIGURE 5.10**

Schematic representation of the progress of the enzyme reaction in liquid-phase pyrosequencing. Light is emitted proportional to the number of nucleotides incorporated. Nucleotides are introduced one at a time; a graph is created that shows the intensity of light and the nucleotide introduced. Enzymes are used to follow the incorporation of nucleotides. *Source: Reproduced, with permission, from Ronaghi[34] (2001).*

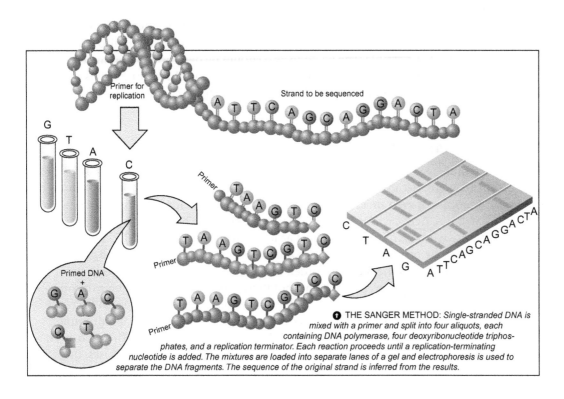

THE SANGER METHOD: *Single-stranded DNA is mixed with a primer and split into four aliquots, each containing DNA polymerase, four deoxyribonucleotide triphosphates, and a replication terminator. Each reaction proceeds until a replication-terminating nucleotide is added. The mixtures are loaded into separate lanes of a gel and electrophoresis is used to separate the DNA fragments. The sequence of the original strand is inferred from the results.*

**FIGURE 5.11**

The basic Sanger method is shown here. In high throughput sequencers, the DNA fragments are made single-stranded then anchored. Fluorescently labeled nucleotides are added, detected, and identified, then the fluorescence is removed. *Source: Reproduced, with permission, from Shaw.*[35]

## 5.10.1 Barcodes

There are two different meanings of barcode in the biological literature. The first, also called unique DNA sequence identifier, refers to a method that enables multiple samples to be pooled for sequencing; each sample is identified by a unique barcode, which enables identification of results during the analysis. The second, called DNA barcoding, refers to a taxonomic method based on a barcode identified within the species' DNA. The discussion here will address only the first meaning.

Unique DNA sequence identifiers are added in a PCR reaction carried out before sequencing, which also adds the primers used for the sequencing reaction. Figure 5.12 shows a diagram

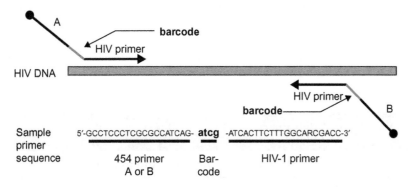

**FIGURE 5.12**

Example of adding unique DNA identifier (barcode) during PCR amplification before sequencing using the 454 (pyrosequencing) system. This step also adds the primer sequence for the 454 PCR. *Source: Reproduced, with permission, from Hoffman et al.*[36] *(2007).*

of a primer used to amplify an HIV gene, before sequencing.[36] To be effective, barcodes must be unambiguous; but because the barcode is part of the sequence read, it is also ideal to have barcodes be as short as possible. The barcode also should be designed to minimize primer-dimer artifacts. Therefore barcode design is a methodological problem, one that is within the purview of the field of bioinformatics (described in Chapter 7).[37] Barcodes are also commercially available.

### 5.10.2 Gene Assembly

Genome assembly refers to the process of putting nucleotide sequence into the correct order. Assembly is required, because sequence read lengths – at least for now – are much shorter than most genomes or even most genes. Genome assembly is made easier by the existence of public databases, freely available on the National Center for Biotechnology Information website (http://www.ncbi.nlm.nih.gov). Just as it is much easier to assemble a picture puzzle if you know what the picture looks like, it is much easier to assemble genes and genomes if you have a good idea of the sequence order. In the human genome, genes occur in the same physical location on the chromosome, but there can be different numbers of copies and variable numbers of repeated sequence that complicate assembly. Although bacterial genomes are much smaller, genes are not necessarily in the same location and multiple copies of the same gene may appear in different locations on the genome. Therefore even with the availability of commercial software and ever growing reference databases, the process of genome assembly can take considerably longer than the time to obtain actual sequence.

## 5.11 HYBRIDIZATION/ANTIGEN–ANTIBODY REACTIONS

The last class of techniques relies on two biological phenomena: hybridization and antigen–antibody reactions (Table 5.6). Hybridization takes advantage of the fact that DNA is double-stranded; when made single-stranded, if the complementary sequence is provided, the strand will bind to the complementary sequence. RNA can be detected by first making cDNA. Antigen–antibody reactions take advantage of the specificity of antibodies for specific antigens. Antibodies are specific to specific antigens (proteins); the usual analogy is to a lock and key. Hybridization reactions are used for DNA and RNA while antigen–antibody reactions are used for protein, but there are some similarities between these two techniques. Both are relatively sensitive and specific measures for identifying a substance, be it genetic sequence or protein. Also, both use the presence of a label to enable detection of the substance of interest (either sequence or antigen) by a detection system.

Just like it sounds, a label is a substance that enables the investigator to find the substrate of interest. There are many different types of labels, ranging from radioactive materials, to materials that change color, or ones that fluoresce when excited by light of a specific wavelength. In hybridization, the label is attached to the probe, a fragment of DNA that complements the sequence of interest. For antigen–antibody reactions the label may be attached directly to the antibody that detects the antigen of interest, or to an additional

**TABLE 5.6** Common Laboratory Techniques Using Hybridization and Antigen–Antibody Reactions

| Hybridization Reactions (Nucleic Acid) | Antigen–Antibody Reactions (Protein) |
|---|---|
| • Southern blotting (DNA) | • Western blotting |
| • Northern blotting (RNA) | • Enzyme-linked immunosorbent assay (ELISA) |
| • Fluorescence in situ hybridization (FISH) | • Immunoprecipitation |
| • DNA microarrays | |

antibody that detects the antibody that is attached to the antigen of interest (a "sandwich" assay). The detection system depends on the label used.

Hybridization and antigen–antibody reactions are extremely powerful techniques and are used in a variety of molecular epidemiologic studies. Hybridization enables the detection of the presence of a specific DNA sequence, such as genes coding for antibiotic resistance or somatic mutation in human cells. With fluorescence in situ hybridization (FISH), the gene can be detected directly in the tissue. Antigen–antibody reactions enable detection of proteins, such as antibodies. Antigen–antibody detection systems that require minimal skill can be created for use during data collection. One commercial product based on this reaction is the over-the-counter pregnancy test.

## 5.11.1 Hybridization

The detection of nucleic acid using hybridization is a versatile way to rapidly detect DNA or RNA with a known sequence. Hybridization makes possible the rapid diagnosis of infectious agents. This is particularly useful for organisms that are difficult to culture or slow to grow, or when rapid detection is important clinically.[38] Hybridization can be used to identify organisms already cultured, or to test a specimen directly. Hybridization enables screening microorganisms for specific genetic characteristics, such as genes for virulence or antibiotic resistance. Gene expression is also detected using hybridization. Hybridization can be conducted in a high throughput format; it is the basis for gene chips, expression arrays, and other types of nucleic acid arrays that are conducted in a micro format that are known as microarrays.

DNA microarrays generally consist of a slide or chip upon which are arrayed minute amounts of (1) DNA fragments, (2) cDNA, or (3) short fragments of nucleotides, known as oligonucleotide polymers, in a $2 \times 2$ matrix. DNA fragments of known genes detect specific genes, cDNA detects gene expression, and, depending on how the oligonucleotides are constructed, oligonucleotide arrays can be used for resequencing, detection of genes, or gene expression. Tens of thousands of DNA fragments or oligonucleotides can be put onto a single array, enabling identification of multiple genes or transcripts in a single experiment. An early application of microarrays was comparative hybridization; this enables direct comparison of an unknown to known sequence, including the genetic sequence of some human pathogens. However, with the development of automated pyrosequencing machines, comparative hybridization has fallen somewhat out of favor as it only detects what is common between organisms and cannot detect what is not on the array. Gene expression arrays, which enable comparison of gene expression between groups, such as diseased and healthy, have a similar limitation, although this is somewhat overcome by using tiling arrays. Tiling arrays use multiple highly overlapping oligonucleotides as probes. In a conventional array, each probe corresponds to a specific gene or gene variant; probes do not overlap in sequence. By contrast, in a tiling array there will be multiple probes that overlap in sequence so that known and unknown variants are identified. For this reason, and because microarrays require more complex analyses, and can have reproducibility problems, some groups prefer differential display (which uses PCR to detect gene transcripts). Tiling arrays can also be useful for resequencing. Resequencing provides greater detail of the sequence for a particular region, detecting minor variations such as SNPs that may be associated with health or disease. Tiling arrays are under development for rapid detection and differentiation of viruses.[40] As the technology improves and the costs fall, microarrays will increasingly be applied in clinical and public health practice, because the potential is high for rapid diagnosis, strain typing, and detection of novel variants.[41]

A different type of microarray is a "Library on a Slide," in which DNA from thousands of bacterial isolates is spotted onto the slide; the slide is then probed with a known sequence.[42] In contrast to other arrays, what is on the slide is unknown. Library on a Slide enables high throughput screening of large bacterial collections for material detectable via hybridization to total genomic DNA.

Like all techniques, hybridization is not perfect. Although microarrays are becoming increasingly standardized, with corresponding decreases in technical variation, there are still variations in hybridization time, temperature, blocking agent, washing procedures, quantity of labeled target, and instrumentation.[43] Arrays intended for diagnosis must be evaluated for sensitivity, specificity, reliability, and reproducibility with the same rigor as any diagnostic test, including assessments within the intended study population.

## 5.11.2 Antigen–Antibody Reactions

Antigen–antibody reactions are a mainstay for the rapid detection of proteins. Antibodies recognize proteins based on their structure as well as content, and can be very specific, binding to only a small part of an antigen (known as the epitope), and discriminating between highly similar epitopes. In nature, if a host is exposed to an antigen, the host will develop an array of antibodies that each bind to a separate epitope of the antigen. These antibodies will vary in specificity. Homogeneous, highly sensitive and specific antibiotics have been developed and propagated in the laboratory. These are known as monoclonal antibodies. Because of their high sensitivity and specificity there are multiple applications for antibodies in clinical and research settings, mostly focused on diagnosis, although there are therapeutic uses, such as passive immunization for botulism, rabies, and other diseases.

Western blotting is a test for protein that shares similarities to Southern and Northern blotting. Like the other blotting methods, the material – here protein – is first separated by gel electrophoresis. The gel is blotted onto a membrane. Then an antibody is used to detect the protein of interest. In a one-step process, the antibody also has a label that is detected; in a two-step process, a second antibody that recognized the first is added. It is the second antibody that is labeled. In addition to research applications, Western blotting is used for confirming HIV infection and diagnosing prion diseases.

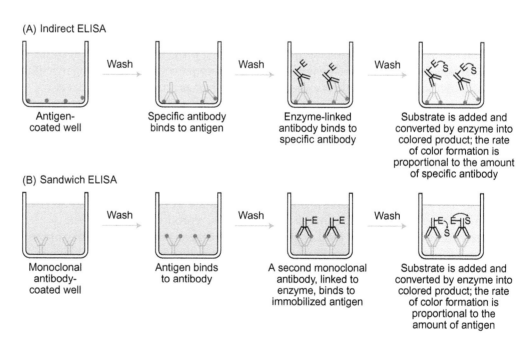

**FIGURE 5.13**

Indirect ELISA and sandwich (direct) ELISA. **A**, In indirect ELISA, the production of color indicates the amount of an antibody to a specific antigen. **B**, In sandwich ELISA, the production of color indicates the quantity of antigen. *Source: Reproduced, with permission, from Goldsby et al.*[39] *(2000).*

The most common application of antigen–antibody reactions is in diagnostics using an enzyme-linked immunosorbent assay (ELISA). For example, ELISA can be used to detect an antibody to HIV; this is known as indirect or sandwich ELISA. ELISA also can detect an antigen for human chorionic gonadotrophin (HCG), which is excreted in the urine of a pregnant woman; this is called direct ELISA. The target (either antibody or antigen) of interest is bound to a substrate. For a pregnancy test, the target is an antibody for HCG, and for an HIV test, an HIV antigen. The target can be on the bottom of a well in a plastic plate or some other substrate. After the test specimen is added, if the protein of interest is present, it will bind to the target. For indirect ELISA, a second antibody is added; this one binds to the specific antibody (Figure 5.13A). For direct ELISA, a second antibody is also added, but this is to the target antigen (Figure 5.13B). The second antibody is labeled with a reporter enzyme that changes color when its substrate is added. The last step is adding this substrate. For the over-the-counter pregnancy test, the entire test takes place on a single platform. As the urine moves through the platform by capillary action, it releases free antibodies and the reporter enzyme. If HCG is present in the urine, it binds to antibodies fixed to the platform; the antibody with a reporter enzyme binds to the bound HCG. Bound HCG is also fixed to the platform as a control. A positive test shows two lines, the control result and that from HCG present in the urine.

Like hybridization assays, antigen–antibody reactions have been put in a microarray format enabling thousands of reactions to occur simultaneously. This makes it possible to rapidly differentiate among infections with similar presentations, differentiate among different bacterial strain types, or screen an individual for past exposure to hundreds of prior infections in a single experiment.

# References

1. B. Alberts, D. Bray, K. Hopkin, et al., Essential Cell Biology, second ed., Garland Science, New York, 2004.

2. P. Stanley, R.D. Cummings, The A, B, and H blood groups, in: A. Varki, R.D. Cummings, J.D. Esko, H.H. Freeze, P. Stanley, C.R. Bertozzi, et al. (Eds.), Essentials of Glycobiology, second ed., Cold Spring Harbor Laboratory Press, Cold Spring Harbor, 2009

3. S. Linden, J. Mahdavi, C. Semino-Mora, et al., Role of ABO secretor status in mucosal innate immunity and H. pylori infection, PLoS Pathog. 4 (1) (2008) e2.

4. Genetics Home Reference: Your Guide to Understanding Genetic Conditions. Insertion Mutation [Internet]. Available at: http://ghr.nlm.nih.gov/handbook/illustrations/mutationtypes?show=insertion (accessed 14.8.10).

5. Genetics Home Reference: Your Guide to Understanding Genetic Conditions. Deletion mutation [Internet]. Available at: http://ghr.nlm.nih.gov/handbook/illustrations/mutationtypes?show=deletion (accessed 14.8.10).

6. Genetics Home Reference: Your Guide to Understanding Genetic Conditions. Frameshift mutation [Internet]. Available at: http://ghr.nlm.nih.gov/handbook/illustrations/frameshift (accessed 14.8.10).

7. S. Talarico, L. Zhang, C.F. Marrs, et al., Mycobacterium tuberculosis PE_PGRS16 and PE_PGRS26 genetic polymorphism among clinical isolates, Tuberculosis (Edinb) 88 (4) (2008) 283–294.

8. D. Mazel, Integrons: agents of bacterial evolution, Nat. Rev. Microbiol. 4 (8) (2006) 608–620.

9. A Basic Introduction to the Science Underlying NCBI Resources. Figure 4. National Center for Biotechnology Information. Available at: http://www.ncbi.nlm.nih.gov/About/primer/genetics_cell.html (accessed 14.8.10).

10. A.A. Brakhage, M. Thon, P. Sprote, et al., Aspects on evolution of fungal beta-lactam biosynthesis gene clusters and recruitment of trans-acting factors, Phytochemistry 70 (15-16) (2009) 1801–1811.

11. O. Johnsborg, L.S. Havarstein, Regulation of natural genetic transformation and acquisition of transforming DNA in Streptococcus pneumoniae, FEMS Microbiol. Rev. 33 (3) (2009) 627–642.

12. M.N. Alekshun, S.B. Levy, Molecular mechanisms of antibacterial multidrug resistance, Cell 128 (6) (2007) 1037–1050.

13. A.D. O'Brien, J.W. Newland, S.F. Miller, et al., Shiga-like toxin-converting phages from Escherichia coli strains that cause hemorrhagic colitis or infantile diarrhea, Science 226 (4675) (1984) 694–696.

14. Molecular Station. Molecular Station [Internet]. Available at: http://www.molecularstation.com/, 2008 (accessed 18.2.10).

15. J. Dale, M.v. Schantz, From Genes to Genomes: Concepts and Applications of DNA Technology, second ed., Wiley, Hoboken, 2007.

16. Huntington's Outreach Project for Education at Stanford. Figure S-2: Gel Electrophoresis. Stanford University. Available at: http://hopes.stanford.edu/n3429/managing-d/genetic-testing (accessed 18.2.10).

17. R.A. Nichols, J.E. Moore, H.V. Smith, A rapid method for extracting oocyst DNA from *Cryptosporidium*-positive human faeces for outbreak investigations, J. Microbiol. Methods 65 (3) (2006) 512–524.

18. Y. Jiang, P.E. Jolly, P. Preko, et al., Aflatoxin-related immune dysfunction in health and in human immunodeficiency virus disease, Clin. Dev. Immunol. 2008 (2008) 790309.

19. J.F. Pissinate, I.T. Gomes, V. Peruhype-Magalhaes, et al., Upgrading the flow-cytometric analysis of anti-Leishmania immunoglobulins for the diagnosis of American tegumentary leishmaniasis, J. Immunol. Methods 336 (2) (2008) 193–202.

20. W.G. Lee, Y.G. Kim, B.G. Chung, et al., Nano/microfluidics for diagnosis of infectious diseases in developing countries, Adv. Drug. Deliv. Rev. (2009).

21. J.E. Fox, An Introduction to MALDI Mass Spectrometry [Internet]. Alta Bioscience: A University of Birmingham business. Available at: http://www.altabioscience.bham.ac.uk/pdfs/Intro_to_MALDI_MASS_SPEC.pdf, 2009 (accessed 14.8.2010).

22. B.C. van Munster, M.J. van Breemen, P.D. Moerland, et al., Proteomic profiling of plasma and serum in elderly patients with delirium, J. Neuropsychiatry Clin. Neurosci. 21 (3) (2009) 284–291.

23. J.M. Matos, F.A. Witzmann, O.W. Cummings, et al., A pilot study of proteomic profiles of human hepatocellular carcinoma in the United States, J. Surg. Res. 155 (2) (2009) 237–243.

24. H.B. Oh, S.O. Kim, C.H. Cha, et al., Identification of hepatitis C virus genotype 6 in Korean patients by analysis of 5′ untranslated region using a matrix assisted laser desorption/ionization time of flight-based assay, restriction fragment mass polymorphism, J. Med. Virol. 80 (10) (2008) 1712–1719.

25. A. Vierstraete, Principle of the PCR [Internet]. Ghent: Ghent University. Available at: http://users.ugent.be/~avierstr/principles/pcr.html, 1999 (Updated 08.11.99; accessed 17.2.10).

26. W.J. Mason, J.S. Blevins, K. Beenken, et al., Multiplex PCR protocol for the diagnosis of staphylococcal infection, J. Clin. Microbiol. 39 (9) (2001) 3332–3338.

27. D.T. McNamara, J.M. Thomson, L.J. Kasehagen, et al., Development of a multiplex PCR-ligase detection reaction assay for diagnosis of infection by the four parasite species causing malaria in humans, J. Clin. Microbiol. 42 (6) (2004) 2403–2410.

28. E.J. Nascimento, U. Braga-Neto, C.E. Calzavara-Silva, et al., Gene expression profiling during early acute febrile stage of dengue infection can predict the disease outcome, PLoS One 4 (11) (2009) e7892.

29. L. Chen, I. Borozan, J. Sun, et al., Cell-type specific gene expression signature in liver underlies response to interferon therapy in chronic hepatitis C infection, Gastroenterology (2009).

30. H. Zhu, M. Liu, P. Sumby, et al., The secreted esterase of group a streptococcus is important for invasive skin infection and dissemination in mice, Infect. Immun. 77 (12) (2009) 5225–5232.

31. K.A. Eckert, T.A. Kunkel, DNA polymerase fidelity and the polymerase chain reaction, PCR Methods Appl. 1 (1) (1991) 17–24.

32. R.M. Ratcliff, G. Chang, T. Kok, et al., Molecular diagnosis of medical viruses, Curr. Issues Mol. Biol. 9 (2) (2007) 87–102.

33. M. Hunt, Real Time PCR. in: Microbiology and Immunology On-line [Internet]. University of South Carolina School of Medicine. Available at: http://pathmicro.med.sc.edu/pcr/realtime-home.htm, 2009 ([updated] 29.10.09; accessed 14.8.10).

34. M. Ronaghi, Pyrosequencing sheds light on DNA sequencing, Genome Research 11 (1) (2001) 3–11.

35. N. Shaw, The Sanger Method. Bloomington: Ned Shaw Studio; Available at: http://www.nedshaw.com.

36. C. Hoffmann, N. Minkah, J. Leipzig, et al., DNA bar coding and pyrosequencing to identify rare HIV drug resistance mutations, Nucleic Acids Res. 35 (13) (2007) e91.

37. P. Parameswaran, R. Jalili, L. Tao, et al., A pyrosequencing-tailored nucleotide barcode design unveils opportunities for large-scale sample multiplexing, Nucleic Acids Res. 35 (19) (2007) e130.

38. M. Klouche, U. Schroder, Rapid methods for diagnosis of bloodstream infections, Clin Chem Lab Med 46 (7) (2008) 888–908.

39. R.A. Goldsby, T.J. Kindt, B.A. Osborne, Kuby Immunology, fourth ed., W. H. Freeman and Company, New York, 2000.

40. Z. Wang, A.P. Malanoski, B. Lin, et al., Resequencing microarray probe design for typing genetically diverse viruses: human rhinoviruses and enteroviruses, BMC Genomics 9 (2008) 577.

41. M. Uttamchandani, J.L. Neo, B.N. Ong, et al., Applications of microarrays in pathogen detection and biodefence, Trends Biotechnol. 27 (1) (2009) 53–61.

42. L. Zhang, U. Reddi, U. Srinivasan, et al., Combining microarray technology and molecular epidemiology to identify genes associated with invasive group B streptococcus, Interdiscip. Perspect. Infect. Dis. 2008 (2008) 314762.

43. A. Brazma, P. Hingamp, J. Quackenbush, et al., Minimum information about a microarray experiment (MIAME)-toward standards for microarray data, Nat. Genet. 29 (4) (2001) 365–371.

# Molecular Tools

There are many molecular tools available and several more under development. The various tools differ by type of substrate that can be measured, DNA, RNA, protein, or metabolites. The type of information that can be gleaned also varies. A test may detect current presence or absence of a substrate, past presence of substrate, and variations in sequence or structure. The tools also vary in validity, reliability, cost, and their amenability to high throughput testing. Many molecular tools facilitate rapid diagnosis, but others give a snapshot of what microbes are present, distinguish among microbes of a single genus and species, or characterize host response.

Molecular tools have opened the door on a whole new set of measures of biological function that complement traditional diagnostic assays and even replace them. Assays detecting DNA, RNA, protein, and metabolites have been developed as diagnostic procedures for a variety of diseases. Genomic tests for the presence of DNA or RNA from an infectious agent enable identification of the cause of a disease within a few hours, in comparison to days or even weeks required for detection using traditional culture techniques. The increased sensitivity of detection has enabled some tests to be conducted on specimens that can be collected with minimal effort, such as saliva, urine, or buccal cells rather than blood or tissue biopsies. DNA-based techniques have entirely replaced culture for the diagnosis of some difficult to culture infectious agents, such as *Chlamydia trachomatis*. Detection of some viral agents, such as human papillomavirus, was not possible before the development of molecular techniques.

The exploration of disease transmission and pathogenesis has been transformed. Identifying the species present is no longer dependent on our ability to grow and characterize the organisms, as trace amounts of DNA and RNA can be used to detect the range of genus and species found in a given environment. This capability enables description and analysis of the interaction of pathogens with the microbiota, microbes that normally inhabit a specific body site, such as in the mouth or gastrointestinal tract. Studies of all genomes present in a specific sample, metagenomic studies, have changed our perception of what organisms live in and on the human body, and where they can live. Using nonculture techniques, presence of infectious agents has been detected in body sites and in environments previously thought sterile or uninhabitable based on culture results. Further, the amounts of each organism can be quantified, giving insight into relationships among organisms. The genomes of infectious agents can be compared to identify genes potentially associated with virulence, and to trace the origin of specific strains or gene regions. Detection of RNA transcripts from a microbe illuminates the order and action of genes associated with pathogenesis; detection of RNA transcripts from the host provides insight into host response to infection. Metatranscriptomic studies help identify how microbes interact with each other and with the human host. Detection of antibodies to bacteria and viruses provides a description of host response to infection, and indicates current or past exposure. Although the promise of interactomics has yet to be realized, studies of posttranslational processing will be essential to understanding the variability of host response to infection and the heterogeneity of transmission, pathogenesis, and virulence of a particular infectious agent, leading to new

Molecular Tools and Infectious Disease Epidemiology.
© 2012 Elsevier Inc. All rights reserved.

therapeutics and prevention strategies. Studies of metabolites, the products produced by cells, can be used to map interactions among infectious agents and between a specific agent and the host. Metabolites can be important markers of disease and disease progression and the effectiveness of therapy.

This chapter first gives a general overview of how molecular tools enhance epidemiology, then the nomenclature of molecular tools – the "omics." The chapter closes with a discussion of choosing the appropriate molecular technique for use in an epidemiologic study.

## 6.1 MOLECULAR TOOLS AND EPIDEMIOLOGY

Molecular tools complement and enhance the various functions of epidemiologic studies: description, hypothesis generation, hypothesis testing, and outbreak investigation. Descriptive epidemiologic studies examine disease patterns in populations to detect emerging threats to public health, develop hypotheses about the underlying etiology, identify possible prevention strategies, and target screening efforts. Descriptive studies also provide information used for planning health services and for guiding clinical diagnosis – by providing estimates of the underlying probabilities of a particular diagnosis in a specific population. Descriptive functions are captured in the epidemiologic triad of person, place, and time. Molecular tools enhance measures of each member of the triad (Table 6.1). Some of the enhancement is in reduced misclassification, or increased sensitivity of detection. Other enhancements are from the added value of using a molecular measure. The number and type of person characteristics that can be measured has been expanded to include indicators of the underlying biology, such as genetic markers. Stored specimens can be tested with a new assay to detect trends over time. Examination of the infecting microbe can

| **TABLE 6.1** Ways Molecular Tools Enhance Epidemiologic Studies, by the Epidemiologic Triads of Person, Place, and Time, and Host, Agent, and Environment | |
|---|---|
| Person | • Move beyond age, gender, race, income, and marital status by characterizing biology of persons who are diseased/not<br>• Enhanced definition of who is diseased/not or exposed/not<br>• Markers for testing hypotheses regarding selection and adaptation |
| Place | • Enhanced identification of reservoir of infection because of increased sensitivity of detection<br>• Because of enhanced definition of who is diseased/not (including subclinical), better able to define place characteristics<br>• Consideration of role of evolution in observed variations by place |
| Time | • Test historical samples to determine trends over time for previously undetected conditions and microbes<br>• Detect order of infection<br>• More accurately determine time of exposure<br>• Determine ancestral origins |
| Host | • Distinguish between factors associated with susceptibility to infection versus susceptibility to disease<br>• Characterize host response to infection, and mediators of that response |
| Agent | • Trace microbial origins<br>• Trace origins of specific genes, such as those associated with transmission or virulence<br>• Detect transmission and virulence mechanisms and mediators of those mechanisms |
| Environment | • Detect exquisitely small amounts in the environment<br>• Discover conditions that enhance or suppress growth of a microbe in a specific environment |

give clues as to the order of transmission, which may not correspond to the temporal order determined by disease detection.

Molecular tools help reveal the underlying biological functions leading to observed disease patterns. Some of the enhancement is in decreased misclassification and increased sensitivity of detection, but a major contribution is the ability to explicitly address the biology at the population level. This also makes it possible to apply principles of evolutionary biology to epidemiology. Like epidemiology, evolutionary biology has a population perspective, and organizes phenomena with respect to person, place, and time. Person variables in an evolutionary sense are gene or epigenomic variants, which can be detected using molecular tools. This enables the generation and testing of hypotheses based on evolutionary theory, such as whether variants are adaptive or undergoing positive selection, which provides an alternative framework for understanding observed phenomena. Similarly, when place variations are interpreted from an evolutionary perspective, if a characteristic occurs over time and space with relatively little variation, this suggests that there is a reason for the maintenance of the factor within the population and that the trait is not new. Finally, both epidemiology and evolutionary biology examine changes over time; but the epidemiologic time frame is generally short (days, months, years) and the evolutionary time frame generally long (generations). The ability to incorporate an evolutionary perspective, and generate and test hypotheses based on evolutionary theory increases the richness and interpretability of epidemiologic data.

In outbreak investigations, epidemiologists use a different triad: host, agent, and environment. The host, agent, environment triad also provides a simple framework for understanding etiology. While more complete (and complex) frameworks have been developed for understanding disease etiology, the host, agent, environment triad gives a simple way to organize the additional contributions of molecular tools to epidemiologic studies. Here too, molecular tools enhance measurement by decreasing misclassification and sensitivity of detection, but also refine measures to better understand the underlying biological relationships. Molecular tools measure host response to infection, enabling us to distinguish between susceptibility to infection and susceptibility to disease. The infecting agent can be described in detail making it possible to determine its origins. Molecular tools enable the characterization of the host and infectious agent interaction, increasing insight into transmission, pathogenesis, and virulence. The ability to detect trace amounts of infectious agents in the environment aids in the identification of new transmission modes and can increase our understanding of the life cycle of a particular infectious agent.

## 6.2 THE "OMICS"

The nomenclature of molecular tools has crystalized to using the suffix "-omics" as a descriptor (Table 6.2). The term genome was coined to be parallel with chromosome, meaning the totality of genetic material from an organism. With the success of the Human Genome Project in sequencing the human genome, additional "omes" were coined, for example, proteome, referring to the totality of proteins from an organism. Hence, the study of genes is genomics, of transcripts, transcriptomics, and so on. Though measuring one ome may provide insight into another, each measure provides a different type of information. Molecular tools facilitate the direct study of genes (genomics), modifications of DNA structure (but not sequence) that can modify or suppress gene expression in a way that can be inherited (epigenomics), gene expression (transcriptomics), proteins (proteomics), metabolites left behind by cell processes (metabolomics), and the interaction of human with nonhuman proteins (interactomics). Metagenomics and metatranscriptomics refer to large-scale cataloging of all genes or gene transcripts in a specific environment. The suffix -omics has spread widely, so this listing is not comprehensive.

**TABLE 6.2** Nomenclature, Definition, and Material Tested of the "Omics"*

| Nomenclature | Definition | Material Tested |
|---|---|---|
| Genomics | The study of genes and noncoding sequences of an organism, and their interactions with each other and the environment | • DNA<br>• RNA |
| Transcriptomics (genome-wide expression profiling) | The study of gene expression in a population of cells in response to different conditions | • mRNA |
| Proteomics | The study of proteins produced by an organism, and their interactions with each other and the environment | • Protein |
| Metabolomics/ biomarkers | The study of chemical fingerprints left behind by cell processes | • Metabolites |
| Epigenomics | The study of changes in gene activity and expression not regulated by genetic sequence | • DNA methylation<br>• Histone<br>• Chromatin |
| Interactomics | The study of the interactions of human and nonhuman proteins with each other and other molecules | • Protein |
| Metagenomics | The study of genomes present in the environment | • DNA<br>• RNA |
| Metatranscriptomics | The study of gene transcripts present in the environment | • mRNA |
| Comparative | Analysis and comparison of "omes" (genomes, proteomes, etc.) from different species | |
| Functional | The study of the function of the components of omes (genes, proteins, metabolites, etc.) | |

*Note: Because the use of some of these "omics" is fairly new, the meaning and interpretation given here reflects what was extant in 2010.*

## 6.3 GENOMICS

The sequence of nucleic acids that determines the genotype (the genetic constitution of an organism) is called genomics. Comparative genomic studies compare the genomes from different species to gain insight into evolution and possible function by analogy. Functional genomics studies observe what happens in an organism when the function of a gene is suppressed (known as knockout studies). By comparing the function of the organism in the presence and absence of gene expression the function of the gene can be deduced. The presence of variations in microbial and animal genomes can be used to trace transmission patterns and the origin of the species (phylogeny). Genetic variations also can be used as markers of pathogenesis, virulence, health, or disease. Detection and identification of new microbes is enhanced by using genomic techniques. Once genetic material is identified, using culture or nonculture techniques, the sequence can be compared to reference databases and the taxonomic classification determined. Techniques such as representational differential display facilitate comparison of genetic sequences present in one tissue but not in another; this technique was used to identify the cause of Kaposi sarcoma.

Molecular epidemiology uses the genetic sequence of microbes and their hosts to describe disease patterns. Description of disease patterns by person, place, and time characteristics provides insight into possible gene function and origin. If a gene variant is associated with disease, it should occur more frequently among individuals with than without disease;

similarly, if it is a microbial gene, it should occur more frequently among microbes causing disease or more virulent disease. If a microbial gene is associated with transmission, it should be found more frequently among strains that are more rapidly transmitted or have a broader geographic distribution. Variations in human and microbial genes affect the person and place characteristics associated with *Vibrio cholerae,* the causative agent for cholera. Persons with the blood group O compared to other blood groups are more susceptible to disease when infected with the El Tor type of *V. cholerae*; epidemic and pandemic cholera has been observed only for cholera serogroups O1 and O139, although other serogroups cause diarrheal disease.[1] (The O serogroup of *V. cholerae* is not related to the human O blood group. The O refers to an antigen on the vibrio cell wall; there are more than 200 known cholera O serotypes.) Of course, interpretation of this type of observation should take into account human population patterns: war, disasters, military recruitment, and schooling can bring together large numbers of susceptible individuals facilitating disease spread.

Micro-RNAs (miRNAs) are RNA molecules that do not code for proteins but instead regulate gene expression.[2] Like ribosomes, miRNAs are also highly evolutionarily conserved and they are found in almost all species, including bacteria and viruses. At this writing, the function of most miRNAs is unknown, but they may provide yet another method for identifying known microbes and detecting new ones.

Characteristics of the microbial genome itself can be informative in explaining disease patterns. RNA viruses tend to be smaller than DNA viruses because RNA is less stable than DNA. RNA viruses also tend to mutate more frequently than DNA viruses, because RNA polymerases lack proofreading ability. The frequency of mutation and genome size of viruses is correlated with transmission mode and disease processes.[3] Although not true for all RNA viruses, RNA viruses tend to be transmitted horizontally between persons by body fluids excreted into the environment or by vectors, and can be highly virulent. The ability to evolve rapidly means that RNA viruses can more easily adapt to a host of a different species. Influenza is an RNA virus spread by aerosols. The duration of infection is very short, from colonization to disease is 1 to 4 days and infection is cleared entirely by 2 weeks.[4] The influenza chromosome is segmented, meaning it is in different pieces, and the pieces can reassort, enabling the influenza to mutate rapidly. Influenza infects birds, pigs, and humans, and when bird flu infects humans, the result is often fatal. DNA viruses evolve more slowly, tend to be transmitted vertically (from mother to child) via person to person close contact, and cause chronic infections. Hepatitis B virus (HBV) is a DNA virus; transmission occurs from mother to child during labor and delivery, via exchange of blood, and close personal contact. The time from infection to disease is long, at 6 weeks to 6 months, and HBV is chronic in 2% to 6% of adults, and up to 90% of infants infected.[5] Chronic HBV infection leads to long-term sequelae; HBV infection greatly increases the risk of liver cancer. These biological differences in duration of infection have epidemiologic implications: RNA viruses generally require larger populations than DNA viruses to maintain transmission. Such insights help explain current and historical disease distributions and transmission patterns, and assist in the prediction of emergence of new infections.

There are also insights to be gained from examining the bacterial genome. Most bacteria have a single circular chromosome, but the presence of extrachromosomal DNA is very common. The ability to acquire extrachromosomal material enables bacteria to rapidly adapt to new environments and ecological niches. Extrachromosomal DNA is acquired via horizontal gene transfer but can be transmitted to offspring. Some extrachromosomal material integrates into the bacterial chromosome, only to separate at a later point. Horizontal gene transfer is responsible for the rapid spread of some mechanisms of antibiotic resistance; the acquired genes for antibiotic resistance can be on a plasmid or integrated into the chromosome. Other virulence factors are also extrachromosomal. The symptoms of diphtheria are from a toxin that is coded by bacteriophage that live in *Corynebacterium diphtheriae*.[6] Similarly, the Shiga toxin found in diarrheagenic *Escherichia coli* O157:H7 is coded by a bacteriophage.

Macroparasites, the fungi, protozoa, and worms are eukaryotes; unlike the prokaryotes (bacteria and archaea) eukaryotes have multiple chromosomes. Because they are more complex, the ability of macroparasites to adapt to different hosts and evolve to resist therapy is considerably slower than either bacteria or virus. Once the ability is developed, however, if there is selection for the characteristic, the ability may spread rapidly. Malaria is caused by a protozoan that is transmitted by a mosquito vector. Quinine, a natural product found in the bark of a tree native to Peru, remains an effective antimalarial; alternatives based on the same chemical were derived, because quinine causes a range of mild to severe side effects that make it less acceptable. Chloroquine, a quinine derivative, was first developed in 1934, and has been used as an antimalarial since the 1940s. By the 1950s, resistance had already been noted.[7]

### 6.3.1 Genomic Techniques for "Fingerprinting" Microbes

Characterizing microbial genotype is the current gold standard for demonstrating the transmission and reservoir of an infectious disease outbreak. Known as "molecular fingerprinting," genomic techniques identify gene variants that can be used to trace the flow of a specific microbial strain through a population, confirm transmission, and relate microbial characteristics to disease manifestations. There are many different fingerprinting methods, which vary in discriminatory ability, ease of conduct, and cost (Table 6.3). The current standards for fingerprinting of many microbes are based on DNA (or RNA for RNA viruses). Genotyping methods look for allelic variations (single nucleotide polymorphisms [SNPs], genetic sequence, pulsed-field gel electrophoresis [PFGE]), variations due to insertions or deletions (restriction fragment length polymorphisms [RFLPs], RFLPs plus Southern blot, PFGE, polymerase chain reaction [PCR]), the presence of DNA palindromic

**TABLE 6.3** Molecular Fingerprint of Infectious Agents

| Name | How It Is Done | What It Measures |
|---|---|---|
| Multilocus sequence typing (MLST) | Polymerase chain reaction (PCR) followed by sequencing of highly conserved regions of the genome | Variations in nucleotide order in selected DNA sequences (MLST is specific to bacteria) |
| Pulsed-field gel electrophoresis (PFGE) | DNA cut with rare cutting restriction enzyme*, followed by electrophoresis | Variation in number of sequences recognized by the restriction enzyme and the length of DNA that occurs between the restriction sites |
| Repetitive elements | PCR, followed by electrophoresis | Variation in number and length of repeated genetic sequences |
| Restriction fragment length polymorphisms (RFLP) | Restriction enzyme, electrophoresis, Southern blotting | Variations in restriction sites, length of DNA that occurs between restriction sites, and insertion sequences |
| Single nucleotide polymorphisms | Various methods to determine changes in genetic sequence | Variations in genetic sequence of specific genes |

*A restriction enzyme cuts the genome at a spot where there is a particular genetic sequence. A rare cutting enzyme is one that recognizes sequence that occurs infrequently across a particular genome.

sequences (sequences that are the same regardless of direction they are read) across the genome (repetitive elements, PCR), or variation in genetic sequence in part or all of the genome (generally limited to virus).

Fingerprinting techniques vary in technical expertise required, rapidity of results, cost, requirements for equipment, and discriminatory ability. New techniques are under development, which increase ease of use and decrease the technical expertise required – although these desirable traits often come at increased cost. Typing tools also vary in discriminatory ability and the number of nontypeables, strains that cannot be resolved into groups. In addition, the applicability of a particular technique in a specific epidemiologic study depends on the characteristics of the particular microbe relative to the research question of interest.

The ease of obtaining genetic sequence is increasing almost as rapidly as the cost per nucleotide identified is falling. Already it is possible to obtain the entire sequence of a bacterial genome that is 5.5 million base pairs, the size of *E. coli*, in just a few days, for less than $1000. Because using genetic sequence to classify microbes is both portable and reproducible, it is likely that as the cost of sequencing continues to fall, the use of methods based on other techniques will be discontinued. In addition to sequencing all or selected regions of the microbial genome, differences in genetic sequence can be detected or inferred by cutting the genome or a single region that has been amplified, and examining the pattern of the resulting fragments on a gel. The DNA on the gel can be transferred to a membrane and probed for the presence of a gene marker, and the resulting pattern analyzed. While currently less expensive than sequencing techniques and requiring less expertise and equipment, gel-based techniques are limited by the difficulty of standardizing procedures and comparing results between laboratories. Another typing technique uses DNA hybridization in a microarray format to identify the presence or absence of genetic markers or genes associated with specific genotypes. Though highly discriminatory, only known variants can be detected using this format. Multilocus sequence typing (MLST), described in Section 6.3.2, is a sequence-based typing technique that is rapidly becoming the gold standard for molecular epidemiologic studies. Other molecular fingerprinting techniques are regularly employed in the United States for surveillance of tuberculosis and food-borne diseases; these are described in greater detail in Section 6.3.3.

### 6.3.2 Multilocus Sequence Typing

MLST is a bacterial molecular fingerprinting method that classifies a bacterial strain based on the genetic sequence of selected genes or gene regions. Similar sequence-based typing methods are under development for viruses that are too large for the entire genome to be sequenced easily. Seven (or more) regions of the bacterial genome are selected, generally from genes that are highly conserved. The regions are sequenced, and the sequence for each region is compared to that in a database. Each allele (variant) observed for a gene region is given a separate number; the allele numbers are concatenated in a set order by gene region to provide a sequence type. A separate MLST schema and associated database must be developed for each species. Currently, MLST is expensive and time-consuming: each gene region must be sequenced in both directions, the sequenced examined for errors, compensated for reference sequence additions or deletions, trimmed to standard length, and analyzed for the correct allele numbers.[8]

*E. coli* is perhaps the best studied microbe, as it is the workhorse of the molecular biology laboratory. *E. coli* is found in most animals, both as a commensal and a pathogen. The MLST for *E. coli* uses the genetic sequence from internal fragments of seven housekeeping genes, which are genes associated with the day-to-day functioning of the organism and whose sequences tend to be conserved because variations can interfere with normal functioning. In the *E. coli* database there are 600 sequence types (STs) that can be grouped into 54 complexes.[9]

Because MLST is based on highly conserved regions, it may be insufficiently discriminatory for use in pathogen tracking or in outbreak investigations. This is true for *E. coli* O157:H7, a major cause of bloody diarrhea and hemolytic uremic syndrome. O157:H7 is highly clonal with nearly identical (>99.9%) nucleotide sequences among different isolates.[10] Since MLST is insufficiently discriminatory for separating *E. coli* O157:H7 isolates from each other, another method such as PFGE or detection of variations in the toxin-coding genes is required.

A system similar to MLST is used in viruses, where the genetic sequence of one or more conserved genes or genes regions is used to classify viral strains into genetic groupings. Hepatitis C virus (HCV) is an RNA virus that mutates rapidly: the estimated mutation rate is ~$2.0 \times 10^{23}$ nucleotide substitutions per site and per year. In theory, there could be a change in every nucleotide every day![11] With a mutation rate this high, there are multiple variants observed within each infected individual. The groups of related viruses found from a single individual are called quasispecies.[12] To compare strains over time and space, the sequence of one or more conserved genes is used as a basis for typing. Eleven different genotypes of HCV have been observed; some are subtypes of the others, so HCV is considered to have six different clades.

### 6.3.3 Gel-Based Fingerprinting

RFLPs are the basis for the gold standard of molecular fingerprinting for tuberculosis. In RFLPs the DNA is cut using a restriction enzyme, an enzyme that cuts the DNA only where it recognizes a specific sequence of nucleotides (a restriction site). The pieces vary in size based on the location of the restriction sites. The pieces are separated using electrophoresis; the gel can be stained and the pattern analyzed. Depending on the fragment size, there can be too many bands to analyze easily. A refinement is to add a Southern blot analysis by transferring the DNA fragments on the gel to a membrane and probing it for the presence a genetic marker. For *Mycobacterium tuberculosis* RFLP typing, the membrane is probed for a transposable insertion element known as IS6110. Tuberculosis genomes vary in number and location of IS6110 (some have none or only one and others as many as 25). When the RFLP is probed for the presence of IS6110, different patterns are observed based on the number of IS6110 elements found in the particular tuberculosis strain, and the size of the particular DNA fragment in which it is found (Figure 6.1). Strains with the same pattern are considered identical; results from fingerprinting might confirm an outbreak or transmission between individuals. Conducting IS6110 RFLP is lengthy, requiring up to five days to complete; it requires considerable expertise and uses large amounts of DNA. Because tuberculosis grows very slowly – incubation up to 8 weeks is required to grow enough to determine tuberculosis is present – RFLP IS6110 does not facilitate rapid answers. Also, although generally quite stable, IS6110 elements are transposable and they have been known move around during storage, complicating strain typing. Further, if the isolates have no or only one IS6110 element, they cannot be resolved easily into groups. Nonetheless, IS6110 typing has played a critical role in increasing our understanding of tuberculosis epidemiology. However, these and other limitations have led the public health community to seek alternative typing methods.

One alternative to IS6110 RFLP typing for tuberculosis is known as MIRU-VNTR, a repetitive element typing

**FIGURE 6.1**

Example of restriction fragment polymorphism probed for the IS6110 element. *Mycobacterium tuberculosis* isolates from routine practice at the Royal Free Hospital. Lanes 4 and 5 have the same patterns, as do 9 and 11. *Source: Reproduced, with permission, from Kanduma et al.[13] (2003).*

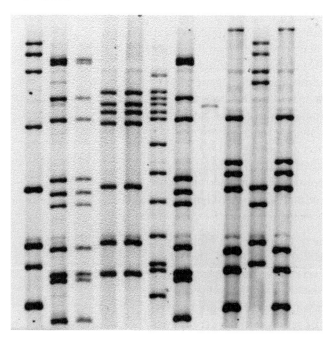

86

method. MIRU stands for mycobacterial interspersed repetitive unit and VNTR, for variable number of tandem repeats. MIRU-VNTR determines the number of copies of MIRU found at several points on the tuberculosis genome. PCR primers are developed based on the sequence flanking the MIRU at each locus. The PCR products vary in size, depending on the number of copies of MIRU present. Like IS6110 RFLP, MIRU-VNTR also has some limitations; one major limitation is that the number of repeats at a particular locus is strongly associated with tuberculosis genetic lineage,[14] thus, choice of loci for inclusion is important for generalizability. However, MIRU-VNTR can be automated, completed in a matter of hours, and the results digitized (Figure 6.2). The rapidity of testing and comparatively low cost are moving MIRU-VNTR to become the standard first method for typing tuberculosis.[13]

Genomic molecular fingerprinting using PFGE is currently part of the Centers for Disease Control and Prevention surveillance system for food-borne and waterborne diseases, known as PulseNet.[15] PulseNet has dramatically increased ability of public health officials to identify and control widely disseminated outbreaks. Recently, the United States Department of Agriculture (USDA) instituted a version of PulseNet called VetNet. VetNet compares PFGE patterns of *Salmonella* isolates from the National Antimicrobial Resistance Monitoring System (NARMS), the USDA, and PulseNet to monitor and identify food-borne *Salmonella* outbreaks.

PFGE is a size-based technique: total genomic DNA is cut into pieces using a restriction enzyme that recognizes a sequence that occurs rarely for the particular microbe. For PFGE the genome usually is cut into 8 to 30 pieces. An electric current is pulsed from different directions to move the relatively large pieces of DNA through the gel; the DNA moves snakelike through the gel. The gel is stained, and the band patterns of the pieces are compared. PFGE is more discriminatory when the chosen restriction enzyme cuts the genome into more rather than fewer pieces. However, more pieces make it more difficult to interpret, especially if some of the bands have little genetic material, making them hard to distinguish from artifact. Some genomes do not have any restriction sites for the standard

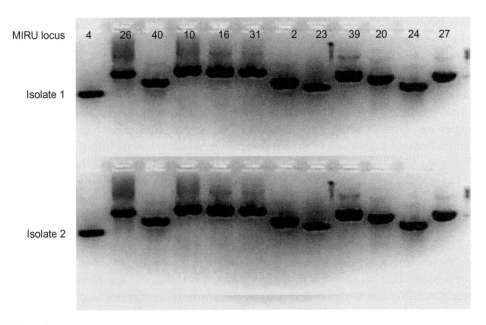

**FIGURE 6.2**

Mycobacterial interspersed repetitive unit (MIRU) of two *Mycobacterium tuberculosis* clinical isolates; in this method several loci where variable numbers of tandem repeats are chosen for replication using PCR. The locus number is shown on the top of each lane. *Source: Reproduced, with permission, from Kanduma et al.[13] (2003).*

enzyme used for a species; cutting with a different enzyme results in a different pattern. Even if an isolate shows the same PFGE type, the type may have occurred by chance alone: only the band sizes are the same, not the genetic content in the bands.

PFGE is also technically difficult and time-consuming and requires special equipment. How isolates run on the gel may vary by lane, although multiple controls are usually included to help in reading. Moreover, there is subjectivity in reading, especially between gels, even if computer software is used. Some bands may be very faint; technical variation may lead to the DNA moving through the gel slightly differently in different lanes on the same gel, and from gel to gel. Despite these limitations, if conditions are optimized, the same restriction enzyme(s) are used, and participants are uniformly trained in the protocols and analytical tools, PFGE patterns are stable and reproducible within and between laboratories.[15] PulseNet requires certification and annual proficiency testing to ensure reproducibility. Because of the complexities, it is best if the isolates to be compared are run on a single gel. The technical limitations make PFGE most informative when there is epidemiologic linkage; such as in an outbreak situation, or when comparing isolates from a single individual over time. In these cases, PFGE supports or refutes the epidemiologic evidence.

### 6.3.4 Resequencing Microarrays: An Emerging Technique for Pathogen Identification and Strain Typing

The human genome and most major human pathogens have been sequenced. The resulting databases provide reference genomes that can be used for pathogen identification and for resequencing. Resequencing refers to sequencing a genetic region where a reference genome is available in order to determine genetic variation. Both pathogen identification and resequencing can be done in a microarray format. The microarray consists of probes that enable pathogen detection and strain variation. The sample DNA is hybridized against the probe on the microarray; depending on the pattern of probes detected in the sample, the pathogen and variant can be detected. Bacteria and viruses can be detected using a single array[16]; other techniques, for example PCR, require separate tests for each suspected pathogen. As the costs of microarrays fall and validity and reliability improve, the microarray format promises to be an extremely useful screening tool for detecting both known and unknown pathogens. Once a class of pathogens is suspected, more specific tests can be developed.

## 6.4 TRANSCRIPTOMICS

The transcriptome measures messenger RNA (mRNA), the message that is translated into protein. Unlike the genome, the transcriptome tells not what genes are present but what genes are currently being transcribed and translated into proteins. It is not an exact correlation, because high levels of mRNA do not necessarily translate into correspondingly high levels of protein production, and high levels of protein production can follow low levels of mRNA[17]; further, some mRNA is not translated to protein, and different forms of the same protein may result from a given transcript.[18] However, at least for now, studying the transcriptome is technically less challenging for many applications than studying the proteome, but this is changing rapidly. When studying microorganisms there are two transcriptomes of interest: that of the host and that of the microorganism. Transcriptomics enables measurement of host gene response to infection, and allows identification of host genes or proteins that are essential for pathogen invasion, growth, or survival.[19] Studying the transcriptome of the microorganism enables identification of genes associated with transmission, pathogenesis, and survival in different environments.

Ubol and colleagues[20] (2008) used a microarray analysis to identify genes expressed among peripheral blood mononuclear cells (PBMCs) from patients with dengue hemorrhagic fever (DHF) to those with dengue fever (DF), that is, a comparison of the transcriptome. Dengue is a viral infection transmitted by mosquitos, also known as breakbone fever. Clinical illness ranges from mild to life-threatening; the most severe forms are DHF and dengue shock syndrome, because they are major causes of morbidity and mortality within countries where dengue is endemic. The reasons for the differences in clinical presentation are unclear, but the study by Ubol and associates[20] showed that PBMCs from patients with DHF compared to PBMCs from patients with DF had higher expression of the genes measured.

Pathogens often express different genes when grown in different environments; what genes are expressed can give insight into pathogenesis. In natural systems, *V. cholerae* is highly infectious, but when grown in vitro it has trouble infecting human volunteers. To explore this phenomenon, the expression profile of *V. cholerae* from stool samples was characterized. The results suggest that infection of humans changes the physiological and behavioral state of the vibrio, causing a hyperinfectious state that contributes to epidemic spread.[21] Another study used transcriptomics to explore hypotheses about why persons with cystic fibrosis are susceptible to infection with *Burkholderia cenocepacia.* It is a major cause of respiratory illness and mortality among cystic fibrosis patients, but does not cause illness in healthy individuals. Persons with cystic fibrosis have a genetic defect which leads to the production of a highly viscous mucus, which is hard for the body to clear from the lungs. This makes cystic fibrosis patients particularly susceptible to respiratory infections. In a sputum model, investigators demonstrated that *B. cenocepacia* alters expression in 10% of its genes when grown in basal salts medium with sputum from patients with cystic fibrosis compared to basal salts medium alone.[22]

## 6.5 PROTEOMICS

Proteomics is the study of the function, structure, expression, and interaction of proteins. Which proteins are expressed depends both on the cell itself and its local environment. Proteins are not merely the translation of DNA into amino acids; the amino acids must be folded into a particular structure to become the active protein. There is often posttranslational processing, such as the addition of carbohydrates or lipids, or the cleavage of parts of the amino acid sequence (gene splicing), making it possible for the same genetic code to result in different protein structures. The sequence of amino acids is known as the primary structure; the final three-dimensional structure is known as the tertiary structure, which has secondary structural components consisting of helices and beta sheets. Studying proteins directly has advantages over both genomics and transcriptomics, as posttranslational processing makes it difficult to predict the final protein structure based on genetic code alone, and studies of transcripts (mRNA) do not accurately predict protein abundance. By studying proteins directly, protein–protein interactions can be identified and quantified, as can the factors that disrupt or modify these interactions. Although studying proteins remains more challenging technically than DNA or RNA, new technologies make this increasingly feasible. It is now relatively straightforward to compare proteins expressed by infectious agents with different characteristics or to describe how an infectious agent impacts host cell functioning. For example, proteomics studies have identified the proteins expressed by *M. tuberculosis* resistant to versus sensitive to isoniazid, and the impact of infection with malaria on the host cell.[18]

There are a number of diagnostic tests based on the presence of specific proteins; these range from measures of specific and innate immunity in the host, to detection of proteins found on the surface of infectious agents. The use of proteins to diagnose disease or to detect past exposure in epidemiologic and clinical applications is not new, but newer tests have increased sensitivity and specificity. For example, pregnancy tests use an enzyme-linked immunoassay to detect the presence of the human chorionic gonadotropin, a hormone released once the embryo implants into the uterine wall. This test is now packaged for home use; this is considerably faster and easier than the initial version from the early 1970s that

**89**

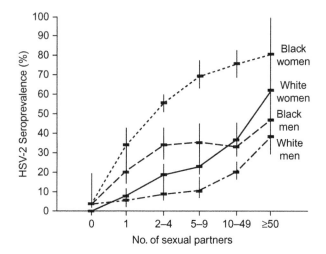

**FIGURE 6.3**

Herpes Simplex virus 2 seroprevalence according to the lifetime number of sexual partners, adjusted for age, for black and white men and women in NHANES III (1988 to 1994). *Source: Reproduced, with permission, from Fleming et al.*[24] *(1997).*

took 2 hours and was conducted in a laboratory.[23] Detecting the presence of antibodies to a specific infectious agent has been used to describe the epidemiology of numerous conditions, using both newly collected and stored samples. Blood samples collected as part of the National Health and Nutrition Examination Survey (NHANES), sponsored by the U.S. government from 1988 to 1994, were screened for the presence of antibodies to herpes simplex virus II.[24] The age-adjusted prevalence increased with lifetime number of sex partners for all groups, with black women having the highest prevalence regardless of number of sex partners (Figure 6.3). These types of analyses help generate hypotheses, evaluate the effects of intervention programs, and aid in prediction of the potential effect of intervention programs.

There are multiple potential applications for applying current and developing proteomic studies in epidemiology (Table 6.4). Already investigators have begun comparing proteins expressed by microbes causing a specific disease to those of the same species that are merely colonizing the body, with the goal of identifying proteins involved in virulence.[25] There are known and yet-to-be-detected markers of disease pathogenesis produced by the host in response to infection, such as antibodies, and markers produced by the infectious agent, such as toxins. The type and number of proteins detectable probably changes with disease stage; describing these proteins will enable detection of disease, disease stage, and past exposure, and a more accurate prediction of prognosis. These in turn will help identify potential vaccine

**TABLE 6.4  Applications of Proteomics in Epidemiology**

- Description of biological processes leading to colonization, pathogenesis, and transmission
- Description of host response to infection, and how the responses mediate disease pathogenesis and transmission
- Improved detection of disease, including intermediate disease stages
- Identification of past exposure to infectious agent(s)
- Prediction of disease prognosis
- Identification of vaccine candidates
- Identification of potential drug targets

candidates and drug targets. As we increase our understanding of the underlying biology, and refine laboratory techniques, the number of applications will only increase. Proteomic studies do have some limitations. Current techniques are insufficient to study rare proteins, ones that are relatively insoluble, or those that are only expressed in vivo.[26]

## 6.6 METABOLOMICS

Metabalomics, also known as metabolic profiling, refers to the study of metabolites, products produced by cells.[27] The products may range from factors secreted by cells to communicate with each other to products resulting from cell processes. By characterizing the metabolites, we gain insight into cell functioning. Identifying metabolite patterns associated with health and disease provides an additional diagnostic tool. These patterns may also give insight into what component(s) of the biological pathway of interest is malfunctioning and hence might be a treatment target.

Metabolic profiles can identify subtle differences between microbial strains that might suggest alternative, more specific therapies. At present, there are no tests that clearly separate uropathogenic *E. coli* from those colonizing the gastrointestinal tract, but a 2009 metabolomic study suggests that it may be possible in the future.[28] This study compared the metabolic profiles of 18 strains of *E. coli* that caused urinary tract infection with paired *E. coli* found in the gastrointestinal tract of the same patients. The *E. coli* that caused urinary tract infection preferentially expressed two siderophores, molecules that scavenge iron, which is something that was not detected by comparing genotype. Metabolite profiles can be used to identify biomarkers of disease or disease prognosis, or the profile itself may be diagnostic. It is early days, but this strategy shows much promise. Preliminary studies suggest that metabolomic profiles of cerebrospinal fluid are sufficient to identify the cause of meningitis without waiting for results of culture.[29] In a mouse model, metabolomic profiles of urine differentiated pneumonia caused by *Streptococcus pneumoniae* from that caused by *Staphylococcus aureus*, and changes in profile were prognostic.[30]

## 6.7 EPIGENOMICS

Changes in gene expression can occur due to changes in the genetic code. However, gene expression can also be regulated by methylation, which does not change the genetic code. The study of changes in gene expression that are not due to changes in sequence is called epigenomics. These changes have been associated with a variety of human diseases, including cancer and development abnormalities, such as Rett syndrome.[31] Analogous changes occur in bacterial, protozoan, and fungal systems.

When a gene can reversibly be switched on and off, and this ability is heritable, the phenomena is called phase variation. Phase variation enables an organism to adapt rapidly to new environments and to evade the host immune response. Though some phase variation is regulated by genomic changes, site-specific recombination, and slippage, other phase variation is regulated by methylation. The expression of the *pap* gene, an important gene for pathogenesis of urinary tract infection caused by *E. coli*, is regulated by methylation (Figure 6.4).

The impact of infection on the human epigenome remains to be described, but this is a potentially fertile area for exploring how infection and other environmental exposures can lead to chronic diseases. Both low and high levels of methylation are associated with a variety of human tumors.[33] Hypermethylation of the *GSTP1* gene, which encodes for an enzyme that defends cells against damage by some caricinogens, occurs frequently among HBV-associated hepatocellular carcinomas.[34]

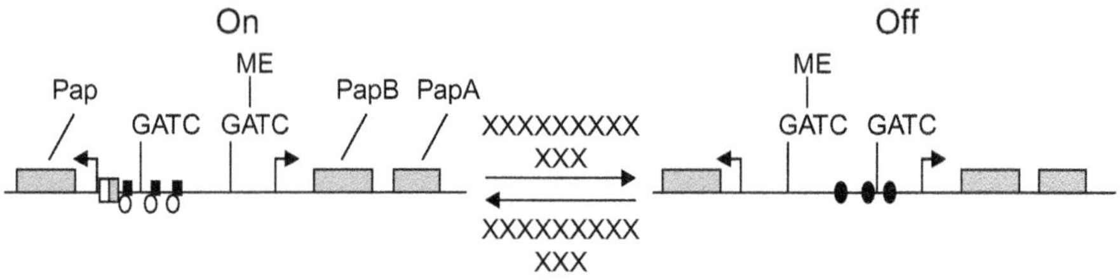

**FIGURE 6.4**

The *pap* locus of uropathogenic *Escherichia coli* contains three genes: PapI, PapA, and PapB, and binding sites for the leucine responsive regulatory protein, GATC. When one GATC site is methylated (ME) by Dam methylase, PapI, PapB, and PapA are expressed; when the other is methylated, they are not expressed. *Source: Reproduced, with permission, from Bayliss[32] (2009).*

## 6.8 INTERACTOMICS

Interactomics refers to the interactions of human and nonhuman proteins with each other and other molecules. Characterizing host–microbe interactions at the molecular level has the potential to identify new measures of host susceptibility, novel strategies to prevent microbial invasion, and develop novel therapies. Many microbes are highly adapted to humans, but within a highly adapted bacterial species there can be much genetic heterogeneity. This is true of *S. aureus*. Perhaps not surprisingly, some host genotypes are associated with carriage of specific *S. aureus* genotypes.[35] Further, there is variation in risk of persistent carriage of *S. aureus*; this variability is associated with certain host inflammatory genes. Similar findings are likely true for other microbes that are both commensals and pathogens. Interactomic studies may shed light on why some individuals are susceptible to recurring infections, and possibly how to block colonization by microbes with genotypes to which they are especially susceptible.

## 6.9 METAGENOMICS AND METATRANSCRIPTOMICS

In the past, technology was insufficient for identifying the community of microorganisms present in the environment or in a specific ecological niche. This is no longer the case. Using PCR it is possible to explore and characterize the genomes and transcripts of organisms found in various ecological niches on and in the human body and in the environment. The exploration of genomes is called metagenomics, and exploration of transcripts is called metatranscriptomics. Current estimates suggest that bacterial cells in and on the human body outnumber human cells by ten to one; if virus, fungi, and archea are counted, human cells are grossly outnumbered. We are just beginning to explore our co-inhabitants. A study of the microbiota on the forearms of six people, which looked only for bacteria, identified 132 species of bacteria – 30 that were previously unknown. Only four species were found on all six people.[36]

Metagenomic and metatranscriptomic studies can be used to characterize the taxonomic groups present, and which genes are functioning. Following DNA or RNA isolation from the specimen, the material can be directly sequenced, made into clone libraries for further testing, or screened for activity.[37] For taxonomic groupings, a universal primer for a highly conserved gene region is chosen that enables taxonomic classification. One highly conserved gene region that has been used for genus identification of prokaryotes and eukaryotes is the DNA that codes for ribosomal RNA (rRNA). Ribosomes use the instructions from mRNA to synthesize protein and are found in all cells. The ribosome consists of two subunits of different sizes (large and small); eukaryotic ribosomes are of different sizes than prokaryotic ribosomes. The sequence of a region of the smaller subunit for bacteria, known as 16S (where S refers to sedimentation rate upon centrifugation) is highly conserved and is often used both for identification and bacterial discovery; for fungi, a region of 18S rRNA is used.

This makes it possible to use a single primer for bacteria or fungi and to identify, at least at the phyla level, the bacteria or fungi present. Primer sets are also available for amplifying viruses. When the transcriptome is isolated, it can be used to create complementary DNA to further identify the genes being expressed. To detect specific gene pathways, primer sets are developed based on reference databases.

There are limitations to using universal primers for detection of classes of microbes. First, there are unknown biases in amplification; undoubtedly some microbes in the class will be missed because the primer is more or less sensitive to the natural variation present. Second, there can be sampling errors; identification of rare microbes can be swamped by those that occur frequently. Third, identification depends on whether the reference database is accurate and complete, and whether variations in genetic sequence correspond to taxonomic groups. Ribosomal RNA is conserved across species that are phenotypically distinct, for example the streptococci, *Streptococcus mitis*, *Streptococcus oralis*, and *S. pneumoniae*,[38] so these species cannot be resolved using rRNA sequence alone. As sequencing technology continues to improve and costs fall, it will be possible to sequence multiple regions of rRNA or other conserved genetic regions to increase resolution.

## 6.10 SELECTING THE CORRECT TECHNIQUE FOR THE RESEARCH QUESTION

Molecular biology is undergoing a period of rapid development; new techniques or modifications of existing techniques appear in the literature almost daily. This poses significant problems for epidemiologic studies, because in epidemiologic studies all data must be comparable. Thus ideally all will be tested using the same technique, if not in the same laboratory. Epidemiologic studies often take place over a number of years. This forces the investigator to decide between storing all specimens for future testing with what becomes the "best" test at the end of data collection or of testing as data are collected with techniques that will be outdated before the study is completed. Specimens might be split so that both paths can be followed, but this may not be fiscally possible. Waiting for the next best test can be problematic. The investigator cannot know the impacts of storage on the validity and reliability of the future test, and also risks losing precious specimens during storage. Current technologies may only be reliable and valid if performed on freshly collected specimens, and future tests may have specific storage requirements. However, using the current best test may make results difficult or impossible to publish several years hence. As tests become outdated it can be difficult to obtain the reagents or kits, making it necessary to determine the validity of the technique with substitute reagents. Thus the investigator is faced with a variety of difficult decisions, upon which hinges the ultimate acceptability of the study results.

A test might be developed specifically for the study. It is possible to develop new detection methods fairly rapidly, for example, a hybridization assay or an enzyme-linked immunosorbent assay. This too can be problematic, because a new test must be assessed for validity and reliability. If the test is for a new syndrome, it may be difficult to determine the validity since classification of the syndrome may be the reason for development. There is also a tendency to use the latest technique when developing a new test; this may or may not be the best choice. The investigator needs to consider trade-offs in cost, ease of use, interpretation, potential for long-term acceptance, reliability, and validity.

As detailed in other chapters, choice of laboratory methods influences study design and vice versa; here, we consider other factors that might influence the investigator's decision-making process (Table 6.5). A test must measure the construct of interest. To evaluate if this is the case, the investigator first needs an understanding of the underlying biology and a well-defined research question. Second, the test must be reliable and valid in the study population. Epidemiologic investigators deal with a range of samples, from both the healthy and the diseased. Commercially available tests are frequently not validated in

---
**TABLE 6.5** Checklist for Choosing Correct Molecular Tool for a Specific Study

- Does the tool measure the construct of interest?
- Are the results valid and reliable?
- Is the tool sufficiently discriminatory in the population of interest? (Is the proportion that cannot be classified low?)
- Do the requirements for specimen collection, storage, and handling fit with other aspects the study protocol?
- Does the expected value of the results equal or outweigh the associated direct and indirect costs?

---

populations similar to the one under study. This often makes it difficult to interpret the findings. Test validity and reliability reflect the populations upon which it was determined *and* the technical capability of those doing the testing. The investigator also should consider the level of discrimination. Discrimination refers to the number of groups obtained during the classification process. The test should be sufficiently discriminatory in the population of interest, and the observed variation should be meaningful. A test that shows little variation has little or no predictive value; if it is highly discriminatory in one subset but not at all in another, results may not provide sufficient information into the biological process to warrant the time, money, and effort involved. Cost is also a consideration, as is the sensitivity to specimen collection, storage, and handling.

For those studying microbes, there are additional concerns associated with the molecular clock. The generation time of microbes is generally considerably shorter than that of humans or other animals. The molecular technique chosen to determine genetic identity may be a different technique from that used to trace transmission within a population. Correctly classifying microbes as part of an outbreak should be insensitive to minor genetic changes that might occur over the course of an outbreak. By contrast, tracing transmission events may depend on identifying the minor genetic changes that occur over the same time period.

## 6.10.1 Measure Phenotype or Genotype?

In molecular epidemiology, molecular tools are used to track a microbe or monitor a biological process. Different tools give different types of information. A major distinction is between genotype and phenotype. Phenotype is generally less discriminatory than genotype, because different genotypes can result in the same phenotype. However, this is not always the case. The correct measure is one that addresses the study goal. If the goal is to detect whether an outbreak of bloody diarrhea was caused by *E. coli* that produces Shiga toxin (STEC), we might use a test for the genes that code for the toxin, grow the bacteria on special media that detects a type of *E. coli* (O157) that produces Shiga toxin, or test for the presence of the toxin in the stool. Any test would address the study goal; the choice would hinge on what test is easily available and is most cost-effective. If we wish to determine whether the same bacteria is causing bloody diarrhea in all temporally linked cases or trace the outbreak back to a specific source, a typing technique that appropriately characterizes the organism beyond the presence of the toxin or toxin genes is needed. However, if we wish to link disease severity or clinical presentation to variations in the toxin, we might choose to look for variations in the genes that code for the toxin, differences in protein structure, or variations in host response to the same organism. Examining the phylogeny of *E. coli* strains to gain insight into the evolutionary history requires a different level of detail, such as examining the genetic sequence at multiple loci around the genome.

Genotype can be characterized in a variety of ways including the presence or absence of a particular gene (something that occurs in bacteria), by variation in the sequence of a particular gene or set of genes, by patterns that are observed when sequence is cut at particular sites, and variants thereof. Some genetic code has no known function, or represents viral or plasmid or insertion sequences that have become incorporated into the genome; these variations can be used to distinguish among microbes of the same species. Phenotype also can be characterized in a variety of ways, including visually recording physical characteristics, such as size or shape, and measuring specific abilities, such as adherence to particular tissues or production of toxins.

Another phenotype used to classify both bacteria and virus is serotype. The serotype (also known as a serovar) of a bacterial or virus surface is determined by the antigens that appear on the surface of the bacterial cell or viral capsule; surface antigens stimulate antibody production. If an animal is exposed to an antigen and produces an antibody specific to that antigen, the antibody will be present in the blood sera. Using blood sera from an exposed animal (or an antibody to the antigen), the antigen can be detected, hence the term serotype. The corresponding genotype is the genetic sequence of the genes that code for the surface antigen. Though some microbes exhibit only a single serotype, others may exhibit more than one serotype. Multiple genotypes may exist within a given serotype. A common serotype makes it possible for the same vaccine to be effective worldwide; this is true for human hepatitis A. However, as different genotypes are found around the world, it is possible to trace outbreaks using genotype.[39] Depending on the microbe, serotype may also be a marker of virulence. There are 90 serotypes known for *S. pneumoniae*, a bacteria that causes pneumonia, otitis media, meningitis, and sepsis, but less than one third of the serotypes are associated with human disease.[40]

## 6.10.2 Requirements for Specimen Collection, Storage, and Handling

Epidemiologic studies are conducted in a variety of settings, from hospitals to street corners. The investigator considers the optimal setting for participant enrollment based on the research question. When comparing settings, the investigator should also consider the constraints imposed by specimen collection, storage, and handling, which are driven by the selected laboratory analyses. It is ideal to get all information and specimens once consent is obtained, because each additional step decreases the chance that the participant will complete the study protocol. This is true even if the added step only requires a short walk down the hall for a blood draw. Without participants and specimens, there is nothing to test.

The feasibility of various specimen collection protocols in the selected study population, and the robustness of the selected test to achievable storage and handling procedures in the setting, should be taken into account when choosing a test. Self-collection might increase participation, but decrease the amount of useable samples. Holding a specimen at room temperature might lead to overgrowth of bacteria, but a preservative might decrease viability of the microbes of interest. Order of processing may facilitate some tests but decrease the validity of others. There is almost always a way to get to a reasonable solution, but it generally takes a lot of thought, time, and preliminary testing.

## 6.10.3 Costs

All budgets are limited. Laboratory tests can be quite costly, and frequently limit the total number of participants that can be enrolled. Therefore the added value of each test relative to the costs should be high. There are direct and indirect costs of testing specimens. Direct costs include costs of supplies, transport, storage, data management, technicians, and equipment. Indirect costs include decrease in study participation and additional institutional oversight from the Institutional Review Board when a specimen is collected. Just collecting the specimen may involve risk to the study participant; the benefits should outweigh the risks. Unlike results of questionnaires, individuals supplying specimens for testing don't know their answers. The investigator has a responsibility to the study participant if test results may lead to early detection of disease; or test results may cause unnecessary worry and suffering. In either case, the investigator should have procedures in place to ethically provide for the participant.

## References

1. D.A. Sack, R.B. Sack, G.B. Nair, et al., Cholera, Lancet 363 (9404) (2004) 223–233.

2. V. Nair, M. Zavolan, Virus-encoded microRNAs: novel regulators of gene expression, Trends Microbiol. 14 (4) (2006) 169–175.

3. E.C. Holmes, Evolutionary history and phylogeography of human viruses, Ann. Rev. Microbiol. 62 (2008) 307–328.

4. A.E. Fiore, D.K. Shay, K. Broder, et al., Prevention and control of influenza: recommendations of the Advisory Committee on Immunization Practices (ACIP), 2008, MMWR Recomm Rep. 57 (RR-7) (2008 Aug 8) 1–60.

5. Division of Viral Hepatitis and National Center for HIV/AIDS, Viral Hepatitis, STD, and TB Prevention, Hepatitis B Information for Health Professionals [Internet], CDC; 2008, Available at: http://www.cdc.gov/hepatitis/HBV/index.htm (Updated 3.12.09; accessed 18.2.10).

6. I. Mokrousov, *Corynebacterium diphtheriae*: genome diversity, population structure and genotyping perspectives, Infect. Genet. Evol. 9 (1) (2009) 1–15.

7. Y. Saito-Nakano, K. Tanabe, K. Kamei, et al., Genetic evidence for *Plasmodium falciparum* resistance to chloroquine and pyrimethamine in Indochina and the Western Pacific between 1984 and 1998, Am. J. Trop. Med. Hyg. 79 (4) (2008) 613–619.

8. S.E. McNamara, U. Srinivasan, L. Zhang, et al., Comparison of probe hybridization array typing to multilocus sequence typing for pathogenic *Escherichia coli*, J. Clin. Microbiol. 47 (3) (2009) 596–602.

9. Environmental Research Institute, University College Cork: MLST Databases at the ERI, University College Cork [website], Available at http://mlst.ucc.ie/mlst/dbs/Ecoli/documents (accessed 8.8.10).

10. A.C. Noller, M.C. McEllistrem, O.C. Stine, et al., Multilocus sequence typing reveals a lack of diversity among *Escherichia coli* O157:H7 isolates that are distinct by pulsed-field gel electrophoresis, J. Clin. Microbiol. 41 (2) (2003) 675–679.

11. H. Myrmel, E. Ulvestad, B. Asjo, The hepatitis C virus enigma, APMIS 117 (5-6) (2009) 427–439.

12. X. Forns, J. Bukh, The molecular biology of hepatitis C virus. Genotypes and quasispecies, Clin. Liver Dis. 3 (4) (1999) 693–716 vii.

13. E. Kanduma, T.D. McHugh, S.H. Gillespie, Molecular methods for *Mycobacterium tuberculosis* strain typing: a users guide, J. Appl. Microbiol. 94 (5) (2003) 781–791.

14. R. Frothingham, Differentiation of strains in *Mycobacterium tuberculosis* complex by DNA sequence polymorphisms, including rapid identification of *M. bovis* BCG, J. Clin. Microbiol. 33 (4) (1995) 840–844.

15. P. Gerner-Smidt, K. Hise, J. Kincaid, et al., PulseNet USA: a five-year update, Foodborne Pathog. Dis. 3 (1) (2006) 9–19.

16. Z. Wang, A.P. Malanoski, B. Lin, et al., Resequencing microarray probe design for typing genetically diverse viruses: human rhinoviruses and enteroviruses, BMC Genomics 9 (2008) 577.

17. S.P. Gygi, Y. Rochon, B.R. Franza, et al., Correlation between protein and mRNA abundance in yeast, Mol. Cell. Biol. 19 (3) (1999) 1720–1730.

18. E.O. List, D.E. Berryman, B. Bower, et al., The use of proteomics to study infectious diseases, Infect. Disord. Drug Targets 8 (1) (2008) 31–45.

19. A. Schwegmann, F. Brombacher, Host-directed drug targeting of factors hijacked by pathogens, Sci. Signal 1 (29) (2008) re8.

20. S. Ubol, P. Masrinoul, J. Chaijaruwanich, et al., Differences in global gene expression in peripheral blood mononuclear cells indicate a significant role of the innate responses in progression of dengue fever but not dengue hemorrhagic fever, J. Infect. Dis. 197 (10) (2008) 1459–1467.

21. D.S. Merrell, S.M. Butler, F. Qadri, et al., Host-induced epidemic spread of the cholera bacterium, Nature 417 (6889) (2002) 642–645.

22. P. Drevinek, M.T. Holden, Z. Ge, et al., Gene expression changes linked to antimicrobial resistance, oxidative stress, iron depletion and retained motility are observed when *Burkholderia cenocepacia* grows in cystic fibrosis sputum, BMC Infect. Dis. 8 (2008) 121.

23. The Office of NIH History: A Thin Blue Line: The History of the Pregnancy Test Kit [website], Available at: http://history.nih.gov/exhibits/thinblueline/timeline.html (accessed 8.8.10).

24. D.T. Fleming, G.M. McQuillan, R.E. Johnson, et al., Herpes simplex virus type 2 in the United States, 1976 to 1994, N. Engl. J. Med. 337 (16) (1997) 1105–1111.

25. M. Ndao, Diagnosis of parasitic diseases: old and new approaches, Interdiscip. Perspect. Infect. Dis. 2009 (2009) 278246.

26. K.L. Seib, G. Dougan, R. Rappuoli, The key role of genomics in modern vaccine and drug design for emerging infectious diseases, PLoS Genet. 5 (10) (2009) e1000612.

27. C.J. Clarke, J.N. Haselden, Metabolic profiling as a tool for understanding mechanisms of toxicity, Toxicol. Pathol. 36 (1) (2008) 140–147.

28. J.P. Henderson, J.R. Crowley, J.S. Pinkner, et al., Quantitative metabolomics reveals an epigenetic blueprint for iron acquisition in uropathogenic *Escherichia coli*, PLoS Pathog. 5 (2) (2009) e1000305.

29. U. Himmelreich, R. Malik, T. Kuhn, et al., Rapid etiological classification of meningitis by NMR spectroscopy based on metabolite profiles and host response, PLoS One 4 (4) (2009) e5328.

30. C.M. Slupsky, A. Cheypesh, D.V. Chao, et al., *Streptococcus pneumoniae* and *Staphylococcus aureus* pneumonia induce distinct metabolic responses, J. Proteome Res. 8 (6) (2009) 3029–3036.

31. A.P. Feinberg, Genome-scale approaches to the epigenetics of common human disease, Virchows Arch. 456 (1) 13–21.

32. C.D. Bayliss, Determinants of phase variation rate and the fitness implications of differing rates for bacterial pathogens and commensals, FEMS Microbiol. Rev. 33 (3) (2009) 504–520.

33. J. Huang, Current progress in epigenetic research for hepatocarcinomagenesis, Sci. China C Life Sci. 52 (1) (2009) 31–42.

34. S. Zhong, M.W. Tang, W. Yeo, et al., Silencing of GSTP1 gene by CpG island DNA hypermethylation in HBV-associated hepatocellular carcinomas, Clin. Cancer Res. 8 (4) (2002) 1087–1092.

35. M. Emonts, A.G. Uitterlinden, J.L. Nouwen, et al., Host polymorphisms in interleukin 4, complement factor H, and C-reactive protein associated with nasal carriage of *Staphylococcus aureus* and occurrence of boils, J. Infect. Dis. 197 (9) (2008) 1244–1253.

36. Z. Gao, C.H. Tseng, Z. Pei, et al., Molecular analysis of human forearm superficial skin bacterial biota, Proc. Natl. Acad. Sci. U.S.A. 104 (8) (2007) 2927–2932.

37. D. Cowan, Q. Meyer, W. Stafford, et al., Metagenomic gene discovery: past, present and future, Trends Biotechnol. 23 (6) (2005) 321–329.

38. J.M. Janda, S.L. Abbott, 16S rRNA gene sequencing for bacterial identification in the diagnostic laboratory: pluses, perils, and pitfalls, J. Clin. Microbiol. 45 (9) (2007) 2761–2764.

39. B. Khanna, J.E. Spelbring, B.L. Innis, et al., Characterization of a genetic variant of human hepatitis A virus, J. Med. Virol. 36 (2) (1992) 118–124.

40. P. Martens, S.W. Worm, B. Lundgren, et al., Serotype-specific mortality from invasive Streptococcus pneumoniae disease revisited, BMC Infect. Dis. 4 (2004) 21.

# CHAPTER 7

# Omics Analyses in Molecular Epidemiologic Studies

The "omics" generate lots of data. Genomics, transcriptomics, and proteomics are testing at high throughput hundreds or thousands of samples at a time or assaying a single sample a hundred or thousand different ways. Ways to analyze, visualize, and compare results, recognize patterns, and assess meaningful differences are under rapid development. The challenge is not only in analysis of the data generated, but in tracking and monitoring the specimens and associated processes used to generate results. The collection and storage of large volumes of specimens requires sophisticated computer software. It is insufficient to have collected and stored specimens; they must be retrievable. Most specimens are frozen. There may be 100 specimens stored in a single box and hundreds of boxes in a single freezer. Computer software enables tracking of each specimen so its location – which freezer, freezer shelf, and box number – is known. Similarly, computer software monitoring ensures testing of each specimen, tracks the results of that testing, and links results to other tests on the same specimen or other data about the study participant.

The ability to rapidly sequence genetic material and the subsequent development of libraries of sequences has led to the development of computer software to assemble genetic sequences, guess what sequence constitutes genes, compare genetic sequences within and across species, and infer evolutionary relationships. Genes can be located on a physical map of a chromosome to distinguish if genes are physically linked, and, with increasing accuracy, the structure of the proteins can be estimated from genetic sequence. Protein sequences can be compared, and predictions made about structure and function. Special tools are available to visualize and analyze gene expression data. Results must be normalized relative to positive and negative controls, the values of each spot determined, duplicate spots compared, and the patterns of results interpreted. This may be the end of analysis or only the beginning; for etiologic inferences or population descriptions these results are linked to clinical, sociodemographic, and behavioral information for further analysis.

This chapter gives an overview of the application of analyses of gene content and gene expression data to molecular epidemiologic studies. The discussion is limited to genomics as these tests and associated analyses are best developed. The intention is to make readers aware of what is possible and familiarize them with some of the associated vocabulary to facilitate communication and creative application of these technologies to population studies. The chapter closes with some comments on the ongoing development of standards for documentation of genetic and other omics data sets.

Molecular Tools and Infectious Disease Epidemiology.
© 2012 Elsevier Inc. All rights reserved.

## 7.1 BIOINFORMATICS, GENETIC SEQUENCES, AND MOLECULAR EPIDEMIOLOGY

Automated sequencing machines and the resulting databases have led to the rapid development of computer software to assemble, manipulate, and analyze genetic sequence. This enables the exploration and characterization of the various genomes and comparisons among them. These explorations can increase our understanding of genetic factors influencing the transmission, evolution, and pathogenesis of infectious agents. The discipline that focuses on the development of ways to use computers to characterize molecular components of living things is bioinformatics. Many aspects of bioinformatics overlap with biostatistics and epidemiology and bioengineering, and readings and coursework on the topics discussed later may be found in each of these disciplines.

The ability to analyze genetic sequences from microbes and humans has transformed the molecular epidemiology of infectious diseases. Whereas the focus of molecular epidemiology in the past was on strain typing to enhance surveillance and outbreak investigation, new phylogenetic methods enable analysis of genetic data to estimate epidemiologic parameters[1] and link epidemic processes to pathogen evolution.[2] Genetic sequences can be compared to identify variants. Genetic sequence can be analyzed to detect the presence of mobile genetic elements, identify potential genes, predict their function, and determine their physical location (Table 7.1). The reader should note that these analyses are all predicated on having databases with previously sequenced genomes for comparison. It is only because sequenced data is made freely available in searchable databases that the field has been able to advance so rapidly.

## 7.2 ASSEMBLE GENE SEQUENCES

Imagine trying to assemble a paragraph of text printed all on one line on a strip of paper that is cut into pieces in the length of five words. Without any additional information this would be a fairly difficult task. Table 7.2 shows this paragraph (with this sentence deleted) cut that way. If you have a template that is very similar – say it only differs in a word here or there – it would be easier. If the number of words could be increased, it would also be somewhat easier. Finally, if there were multiple copies of the paragraph cut into five-word pieces with a random start it would be easier still. This is essentially how a genetic sequence is assembled.

| TABLE 7.1 Types of Genetic Analyses and Their Application in Molecular Epidemiology | |
| --- | --- |
| **Technique** | **Application in Molecular Epidemiology** |
| Compare and analyze genetic sequences | • Assemble genetic sequences<br>• Identify single nucleotide polymorphisms (SNPs), plasmids, phage, transposable elements, and other gene variants<br>• Detect evidence for recombination, selection, saturation<br>• Identify parsimonious informative sites that can be used to trace outbreaks or determine phylogeny<br>• Assess genetic diversity<br>• Identify potential genes<br>• Identity of genes to homologues across species<br>• Predict gene function and protein structure<br>• Detect pathogenicity islands<br>• Determine genetic lineage |
| Determine physical location of genetic sequence on the chromosome | • Distinguish between functional and physical linkage |

**TABLE 7.2** First Paragraph of Section 7.2 Cut Into Five-Word Fragments and Sorted by Letter of Each Fragment. Notice That There Are Several Ways to Assemble the Fragments That Make Possible Sense

a fairly difficult task. If
a random start it would
additional information this would be
copies of the paragraph cut
differs in a word here
easier. If the number of
Finally, if there were multiple
genetic sequence is assembled.
Image trying to assemble a paragraph
into five-word pieces with
into pieces in the length
is very similar – say only
of five words. Without any
of paper that is cut
of text printed all on
one line on a strip
or there – it would be
start it would be easier
still. This is essentially how
would also be somewhat easier.
words could be increased, it
you have a template that

There are several ways that the fragments in Table 7.2 might be put together into sentences to make apparent sense. Unlike genetic sequences, there is only one deletion, and it is an entire sentence, with no insertions, inversions, duplications, and repeated elements that make sequences hard to assemble, even with a template present. The fragments in genetic-speak are known as contigs, which is short for contiguous overlapping sequence. Current sequencing technologies result in contigs of lengths from 500 to 1000 base pairs for Sanger sequencing. These lengths are rapidly being approached by methods sequencing by synthesis (pyrosequencing). Moreover, new methods are under development to increase read lengths, because longer contigs are easier to assemble. Even with longer contigs, having a scaffold or template increases ease of assembly. Genomes are being sequenced daily and deposited in databases. Once deposited, these sequences can be used to construct a scaffold for assembly. This makes the next generation of sequencing methods and microarrays for resequencing with a high degree of resolution increasingly feasible.

Now consider some actual genetic data, part of the genetic sequence of a heat shock protein found in *Chlamydia pneumoniae* (Table 7.3). Even in this short fragment there are two sequences of seven nucleotides that are identical that might cause confusion in assembly (gtaaaag, shown in bold). Insertions, inversions, duplications, single nucleotide polymorphisms (SNPs), and repeated elements are of particular interest to molecular epidemiologists as they are a source of variation that can be markers for or the cause of the factor of interest.

**TABLE 7.3** Part of the Genetic Sequence of *Chlamydia pneumoniae* 600-kD Heat Shock Protein GroEL (groEL) Gene [GenBank: AF109791.1]

cttgcagaag ca**gtaaaag**t tactctaggt cctaaaggac gtcacgtagt tatagataag
agctttggct ctccccaagt gactaaagat ggtgttactg tagctaaaga aatcgagctc
gaagacaaac atgaaaacat gggcgctcag atg**gtaaaag** aagtcgccag caaaactgct

## 7.3 COMPARE AND ANALYZE GENETIC SEQUENCE

A great deal can be learned from comparing and analyzing genetic sequences. Comments can be made about variation in genetic content within a species, potential genes detected, guesses made about possible gene function and protein structure, and the origin of specific genes and evolutionary relationships. A first step in any comparison is ensuring that the correct sequences are being compared, which involves aligning the sequences. Aligning is essentially laying out the sequences for comparison in rows and comparing nucleotide by nucleotide (for nucleic acids), which can be complicated when there are inversions, insertions, and deletions or recombinants (recombinants are pieces of genetic material acquired from another genome).

Unlike the human genome, microbial genomes can be quite plastic. Some genomes have high mutation rates, such as the retrovirus HIV, which causes AIDS. In HIV, mutation occurs during replication: an error is introduced in approximately 1 out of every 2000 nucleotides incorporated. Others, like the bacteria *Escherichia coli*, which is a common inhabitant of animal intestinal tracts and is also capable of causing a range of diseases from diarrhea to pneumonia, the spontaneous mutation rate is $\sim10^{-6}$ per gene per generation. However, the *E. coli* genome is very heterogeneous; large pieces of DNA can be acquired through horizontal gene exchange. The *E. coli* K12 strain is 4.6 million base pairs compared to the 5.5 million base pairs of *E. coli* O157:H7. K12 was first isolated in the 1920s and has been used extensively for a variety of studies; O157:H7 is a Shiga toxin–producing strain that causes diarrhea and has severe complications, including hemolytic uremic syndrome. The difference in genome size is partly due to the inclusion of phage genomes encoding Shiga toxin.[3] (Bacteriophage, known as phage, are viruses that infect bacteria.) Shiga toxin was first identified in *Shigella*, another species of bacteria that also causes diarrhea. Not all genomes are as heterogeneous as *E. coli*, but others such as *Haemophilus influenzae* that causes otitis media (ear infections) are even more so.

Identifying variation in genetic content provides an important marker for molecular epidemiologic studies. Multilocus sequence typing (MLST) and other sequence-based methods are becoming standard fingerprinting methods (see Chapter 6). Genetic variation can be used to differentiate between outbreak and nonoutbreak strains; if a pathogen has a high mutation rate, variation can be used to trace transmission. Genomic variants can also be markers or causes of differences observed in transmission or pathogenesis. Comparing the number and type of deletions and insertions (known as indels), inversions, SNPs, and duplications between isolates from commensal and invasive strains or strains causing different clinical presentations can help identify more virulent or transmittable variants and point to the potential reason(s) for those differences.

Roumagnac and colleagues[4] investigated the evolutionary history and the population structure of *Salmonella enterica* serovar Typhi using 200 gene fragments ($\sim500$ base pairs) among 105 strains collected worldwide. The genes selected included 121 housekeeping genes. Surprisingly for an obligate pathogen, the analysis of the housekeeping genes suggested a genome not under strong selection; the population structure was consistent with neutral selection or genetic drift. Typhi has an asymptomatic carrier state that occurs in some individuals; the authors suggest that the carrier state enabled persistence in isolated groups. They also found haplotypes persisting over a period of 44 years. By contrast, Typhi is under strong selection by antibiotics. They observed clonal expansion of resistant strains, with outbreaks of resistant strains leading to transmission between countries and continents. However, antibiotic-resistant clones did not completely replace sensitive strains, again underlying the importance of the carrier state for maintaining transmission.

Examining the genome is also informative. The genetic code can be read to identify open reading frames (ORFs), potential gene coding regions. Using public databases, ORFs can be compared to genes in other genomes to see if it is similar (homologous) to a known coding sequence (gene). Public databases annotate, that is, attach biological information to a specific gene, which can give insight into function. Further, the structure of the protein that the DNA codes for can be hypothesized, which provides further insights. Genome analyses are useful for generating potential explanations for why an ORF or variants of an ORF or gene might occur more frequently among more virulent pathogens, for example.

Different species have different proportions of guanine + cytosine (GC). Regions within a genome where the GC proportion is markedly different from the whole may have been acquired via horizontal gene transfer. This insight has opened a new area of research that is investigating circumstances that enhance horizontal gene exchange. One example of such a region, known as pathogenicity islands, are horizontally acquired groups of genes that contain virulence factors. Antibiotic resistance is often exchanged among bacteria of the same and different species via horizontal gene transfer. Understanding the emergence and transmission of virulence and resistance genes via horizontal gene transfer will help generate alternative strategies to slow or prevent their emergence and transmission.

An analysis of eight strains of *Streptococcus agalactiae* representing different serotypes illustrates the utility of genomic analysis.[5] *S. agalactiae* has nine known serotypes. The different serotypes have been associated with different potentials for virulence. Genomic analysis confirmed what had been suggested by MLST comparison, that genetic content does not cluster by serotype: there was greater identity between sequenced strains from very different serotypes than between two strains of related serotype. The core genome shared by all isolates was ~80% of any single genome; the remaining 20% of strain-specific genes included 69 genomic islands. An analysis of the genetic sequence suggested that the genetic islands may have been acquired by horizontal gene transfer. Phage-associated genes accounted for 10% of strain-specific genes. The most striking finding was that the gene reservoir for *S. agalactiae* is so large that the authors predicted new genes will continue to be discovered even after hundreds of genomes have been sequenced.

## 7.4 GENE MAPPING

Identifying gene location enables further study and manipulation of specific genes. If genes are close together on the genome they are more likely to move together. A genetic marker, such as a SNP, may be associated with a particular disease or function because it is located near the actual gene of interest. On the other hand, some genes share a common function, so even if they are not located physically close together, they tend to occur together. The human genome has genes in fixed physical locations but bacterial genomes often do not. Gene location is extremely helpful for polymerase chain reaction (PCR) primer design. Having genetic sequence and a map makes it possible to design primers to cover a region of interest. For example, if the purpose is to look for variations within an entire gene, multiple primers might be developed (depending on the size of the gene) that begin both inside and outside the gene region. In bacterial genomes, genes may be present or absent or the genetic code may be variable or degenerate or truncated. Thus it is often useful to design primers outside the region in case there are variations that limit the effectiveness of primers based on the recorded gene sequence.

In all species there are gene regions more prone to variation than others. Because of selective pressure some genes are highly variable, particularly those encoding cell surface proteins. Identifying the location of the genetic region or gene makes it possible to study it specifically. Because physical location matters in both conducting laboratory work and interpreting experimental findings, genetic maps are extremely useful. Genomes can be very large, and vary in content, so there are two different types of genome maps: genetic maps that give landmark locations or markers, and physical maps that show both the markers and the genetic sequence between the landmarks. Relative location can be determined by examining genetic sequence available in public databases for the presence of restriction sites or other markers. Knowing a gene location enables studies to determine the exact gene sequence, gene function, and gene variation. Biological information (annotation) associated with markers in the same physical region can give additional information regarding gene function.

Bacterial genomes of the same species (by phenotype) can vary wildly in genetic content, with comparisons of two strains of the same species having less than half of the genes in common. Taxonomy is difficult under these circumstances. In recognition of this genetic variability, bacterial genomes are characterized by their core genome, the number of genes found in all members of the species, and the pan-genome, the global gene repertoire of a bacterial species.[5] In an analysis of 20 sequenced E. coli strains, the core genome was only 42% of the genes (1976 genes out of an average of 4721 genes); the pan-genome contained 17,838 genes.[6]

## 7.5 BIOINFORMATICS, MICROARRAYS, AND APPLICATION TO MOLECULAR EPIDEMIOLOGY

Microarrays make it possible to conduct thousands of experiments simultaneously, either interrogating a single specimen a thousand different ways or a thousand specimens a single way. The experiments are based on hybridization or antigen–antibody reactions. Arrays may include nucleic acids, antibodies, or other proteins, or small molecules (peptides, carbohydrates) (Figure 7.1).[7] Some arrays contain the reference material, such as oligonucleotides representing genetic sequence; others contain the unknowns, such as tissue or bacterial DNA. If the array contains reference material it is possible to

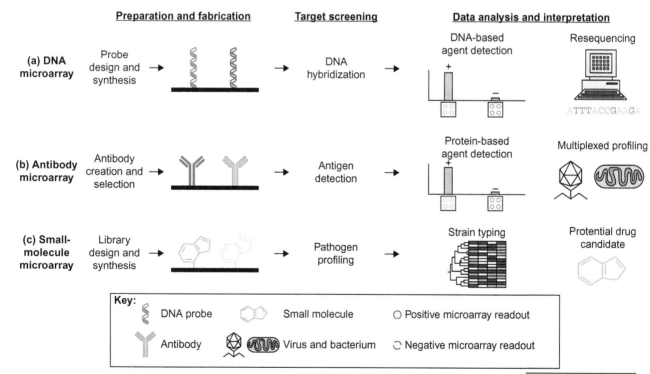

*Trends in Biotechnology*

### FIGURE 7.1

DNA, antibody, and small molecular microarrays. The microarrays differ in the substrate bound to the surface and the detection principle. DNA microarrays are based on hybridization, antibody on antigen–antibody reactions, and small molecular arrays on antigen–antibody reaction and other binding reactions. *Source: Reproduced, with permission, from Uttamchandani et al.[7] (2009).*

interrogate one specimen a thousand different ways. If the array contains unknowns, thousands of different specimens can be interrogated at a time. Arrays of nucleic acids may contain sequence of complementary DNA (cDNA) to detect gene expression (RNA) or oligonucleotides to compare the sequence with reference genome(s) arrayed on the slide or for resequencing. To detect proteins, the array is composed of antigens or antibodies. Small molecules can be arrayed and screened with pathogens, facilitating identification of potential drug targets.

The large amount of data generated requires substantial data management and manipulation even before the results can be interpreted. With thousands of spots on an array as small as an inch square, and each spot corresponding to an experiment, it is essential to be able to be able to identify which specimen is in which spot or which experiment corresponds to which spot (Table 7.4). Spots must be normalized to reference values, because the amount of signal can vary with the amount of target in the spot, and there is variation between runs. Comparisons to reference values, control positive and negative spots, are required to appropriately interpret a result. Most arrays include replicates of each spot; the replicates must be analyzed for agreement. If the values do not agree, the investigator must decide on whether to repeat the test using a same or different method. Once it is certain what the values are, data must be interpreted.

Microarrays are yet to be a "plug and play" item. Like any test, there is variability from run to run; replicate experiments should be run to assess reliability. The analysis is complex and beyond the scope of this book, but involves normalizing results relative to positive and negative controls before any analysis of the results per se. Commercial products use different platforms that generally have their own computer software. Analysis focuses on recognizing patterns associated with particular outcomes. Because microarrays are still expensive, the sample sizes of most published studies are small. This exacerbates problems of detecting significant results by chance alone. For arrays that conduct thousands of experiments simultaneously on a single sample, the potential to detect effects by chance alone is high: with an alpha level of 5% in 1000 tests we would expect 50 to be statistically significant by chance alone. Therefore it is essential to estimate the rate of false positives, and adjust for this rate in the analysis. Several statistical methods are available.[8]

Commercial microarrays are increasingly available for research purposes. The PhyloChip (PhyloTech, Inc., San Francisco, California) is an oligonucleotide microarray that uses the 16S rRNA gene to identify microbes. The chip enables rapid identification of multiple bacterial and archaeal species from a medical or environmental sample.[9] Another oligonucleotide array, ViroChip (University of California, San Francisco), targets viruses. The array includes the most highly conserved 70-mers from each viral family, which enables detection of known and unknown viruses.[10] It is also possible to buy custom arrays, that is, arrays with the genetic sequences of interest, such as arrays that include the sequences of several bacterial genomes or arrays with antibodies that detect human cytokines.

105

**TABLE 7.4** Bioinformatics, Microarrays, and Applications to Molecular Epidemiology

| Technique | Application in Molecular Epidemiology |
|---|---|
| Mapping spots on an array | • Linking results to other biological and sociodemographic data |
| Normalizing results relative to positive and negative controls | • Accurately interpreting findings |
| Identify patterns | • Gene expression profiles used to group by disease or exposure status |
| Resequencing | • Strain typing<br>• Detection of novel strains or genetic variants |

As more genomes are sequenced, microarrays become increasing useful for resequencing. Resequencing refers to sequencing a genetic region for which a reference sequence is already known to identify variants. Microarrays provide an efficient platform for resequencing; fine-tiling arrays use overlapping probes with variants so SNPs can be detected. When applied to pathogens, resequencing arrays enable rapid strain typing and detection of new strains and novel variants. High-density resequencing arrays have been proposed for rapid identification of reassorted influenza A strains.[11] Similarly, when applied to the human host, resequencing arrays enable detection of new gene variants and variants associated with increased or decreased risk of disease.

Investigators can also create their own custom arrays. Large bacterial collections might be arrayed on a slide to create a library on a slide where each spot represents a single bacterial genome.[12] Similarly, investigators might create tissue arrays where each spot is from a different specimen. Creating an array in-house poses particular problems. If arrays are created using complementary DNA (cDNA), large numbers of PCR reactions must be conducted, the products must be checked, and then they are spotted on the array. If creating a library on a slide, DNA must be extracted, checked, and then spotted on the array. Because hybridization can vary across the surface of an array, positive and negative controls should be distributed accordingly. If creating a library on a slide or tissue array, specimens should be distributed randomly according to known characteristics, for example, disease or exposure status. It also is prudent to include replicates of each sample on the array, again distributed randomly, so any within-array variation can be taken into account during the analysis. Creating arrays requires special equipment, at a minimum an arrayer, but it is extremely useful to have a robot to handle processing steps either of the cDNA or specimens. Robots and arrayers come with custom software so that the investigator can determine the location of each specimen/cDNA on the array.

A limitation of the microarray format is that it only can detect what is present on the array. For example, imagine that a DNA array contains genes representing the genomes of three *E. coli* strains that have been sequenced. The investigator conducts a series of comparative hybridization experiments using a range of *E. coli*. Since *E. coli* has a very heterogeneous genome, it is likely that the genomes of the tester *E. coli* might share as little as 30% of the genes found on the slide. The experiments will identify what genes are shared and what known genes are not shared, but cannot detect genes present in the tester *E. coli* not present on the slide. However, as increasing numbers of genomes are sequenced and can be included on an array, this becomes less of a disadvantage.

The promise of microarrays for molecular epidemiologic studies of infectious diseases is twofold. First, microarrays provide an efficient platform for exploring host–agent interaction. There clearly is genetic variation in host response to infection, even with similar doses of the same strain of a pathogen. Understanding the source of this variation should ultimately lead to strategies to minimize the adverse effects, such as rare serious sequelae or tendencies to be highly infectious. Microarrays also enable rapid screening of host response to infectious proteins. The resulting rapid identification of proteins that stimulate long-term host protection can dramatically shorten the time required to develop new vaccines, and is the basis for the emerging field of reverse vaccinology.[13] The expectation is that microarrays will advance our understanding of various aspects of the host–agent interaction, resulting in novel, more accurate, and rapid diagnostic and prognostic tests, new vaccines, and novel therapies.

Second, microarrays have great potential for increasing our understanding of pathogens. Microarrays can characterize the pathogen response to infection, that is, what genes are expressed during different stages of infection, from colonization to invasion. As genes associated with virulence are identified, using microarrays or other methods, population-based collections of pathogens can be screened to determine gene prevalence to identify the most common variants, as these are best candidates for intervention. Infectious agents can rapidly be screened for sensitivity to a large number of drug targets using small molecular arrays. As we increase our understanding of infectious agents, microarrays can be used for rapid diagnosis and strain typing. Since the amount of material required is small, and the agent need not be cultured, this promises to be an important advance for surveillance of slow growing or rare infections, such as tuberculosis, or of viruses, such as influenza, where multiple viral agents can cause similar clinical presentation.[7] In principle a single microarray could identify the infectious agent, its resistance to known therapies, whether any known virulence genes are present, type the strain, and identify new variants.

## 7.6 DETERMINING SIMILARITY AND RELATEDNESS

The ability to determine the similarity among microbes or to determine evolutionary relationships has multiple applications in molecular epidemiology (Table 7.5). When studying an infectious disease outbreak, or tracing an epidemic, pathogens causing the outbreak and epidemic must be differentiated from nonoutbreak or epidemic strains. As in epidemiology in general, this classification is based on markers, such as a particular phenotype or genotype, which may have no biological association with the outbreak itself. Microbes are populations; with whole genome sequencing it is apparent that even during outbreaks there are minor variations with the pathogen. Indeed, PCR of the population of variants results in a consensus sequence.[14] To trace outbreaks or confirm transmission it is essential to have a way to determine if any observed variations are consistent with those that might be expected over the course of an outbreak.

For other epidemiologic applications, it is useful to analyze variations in genetic sequence to identify evolutionary relationships. Variations in genetic sequence occur due to mutations during replication (bacteria can also acquire genes via horizontal gene transfer). The rate that mutations occur in genetic regions not under selection can be used as a molecular clock to give insight into the timing. Using the molecular clock it is possible to estimate when an infectious agent jumped species or (for fast-evolving organisms) to establish order of transmission. The rate of mutation varies by pathogen; knowing the underlying mutation rate is essential for accurately interpreting strain typing results. These variations can enhance analysis, by enabling tracing of the order of transmission; or they can detract if they lead the investigator to erroneously assume a variant is not an outbreak strain.

To group strains based on genetic variations, epidemiologists have turned to the tools of evolutionary biology, which has developed several analytical techniques for determining the similarity and relatedness. Some of the methods used to analyze genetic data are hierarchical

107

**TABLE 7.5** Determining Similarity and Relatedness

| Technique | Application in Molecular Epidemiology |
| --- | --- |
| Determine similarity | • Trace outbreaks<br>• Confirm transmission |
| Determine evolutionary relationships among specimens and ancestral origin | • Identify ongoing epidemics<br>• Trace reservoir of infection<br>• Track emergence and spread<br>• Identify transmission patterns<br>• Trace source of antimicrobial resistance or virulence<br>• Identify lineages associated with disease<br>• Estimate $R_0$ (reproductive number)<br>• Estimate population size required for persistence<br>• Predict course of ongoing epidemic and emergence of new variants |

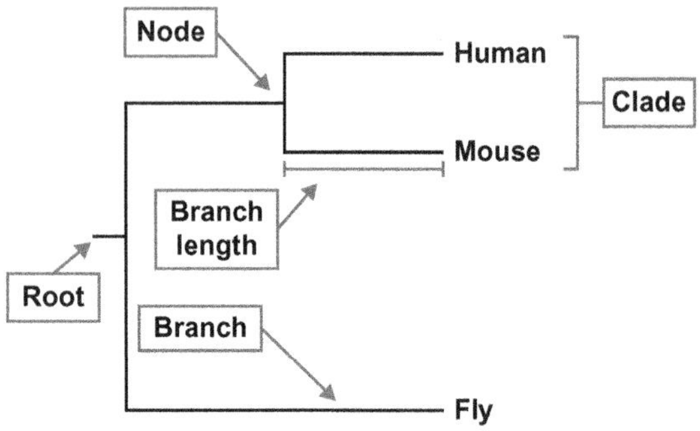

**FIGURE 7.2**
Nomenclature of a dendrogram. *Source: Reproduced, with permission, from the National Center for Biotechnology.*[15]

clustering algorithms (unweighted pair group method using arithmetic averages [UPGAMA]) and similarity indices (Dice, Sorenson's pairwise coefficient). These metrics are applied to results from gel-based methods such as pulsed-field gel electrophoresis (PFGE) and restriction fragment length polymorphism (RFLP), as well as genetic sequence. The calculation of these and other indices is the subject of several textbooks and will not be presented here. Clustering algorithms enable grouping, and similarity indices indicate how similar the groups are to each other, based on some metric (such as genetic distance).

The relationships among units and their indices of similarity can be represented graphically using a dendrogram or tree (Figure 7.2). A tree provides a simple way of illustrating relationships, and identifying clusters of closely related genotypes that can be used in epidemiologic analyses. The tree can be rooted or unrooted. If rooted, the root represents the common ancestor of all the taxa in the tree. Phylogenetic analysis was developed to estimate order of descent. Because we cannot observe evolutionary history – only what is present now – phylogenetic trees are guesses about past relationships. A rooted tree has an implied order of descent, read from the root outwards. In addition to defining ancestor–descendent relationships, a rooted tree is used to estimate the evolutionary distance and the time in the past where two strains or sequences or species diverged. The branches show the relationship between the taxa in the tree. The length of the branch, in a scaled tree, represents the number of changes that have occurred between the taxa at the tip of the branch and the node. A clade is a group of two or more taxa or DNA sequences that includes both their common ancestor and all of their descendants. Although phylogenetics was developed using species as a unit of analysis, in molecular biology the unit of analysis may be another unit, such as bacterial strain.

Trees can be drawn in different formats. In Figure 7.3 the figures in the first two rows are the same tree shown in two formats. In the first row, the branches of rooted trees are unscaled, meaning branches are not proportional to evolutionary distance. The nodes represent when the two species diverged on a time scale. The time to the common ancestor of species A and B is the same, as it is for C and D. The time to the common ancestor of A, B, and E is the same, but further in the past than the common ancestor of A, B, C, and D. Time can be estimated using a molecular clock, the fossil record, or stored samples. The later is frequently possible for microbes. In the second row is a rooted tree with scaled branches. The scale represents the number of changes between the sequences in the units shown. Changes can be evolutionary changes, genetic sequence changes, or some other type of changes. The common ancestor of species A and B is further away for species A than species B. The distance between A and B is the sum of the branch length from where A and B diverge to A plus the branch length from where A and B diverge to B. Unrooted trees, shown in the bottom row, make no assumption regarding a common ancestor. Scaled branches in an unrooted tree still mean the number of evolutionary changes or sequence changes. Unrooted trees are often used in outbreak investigations.

An analysis of isolates analyzed as part of a case-control study of a food-borne outbreak caused by *S. enterica* serotype Montevideo in soft, raw-milk cheese demonstrates the utility of these analyses (Figure 7.4). Here PFGE patterns of DNA from case-patients, the putative source and nonoutbreak associated patients are displayed as a phylogenic tree. This scale is unrooted, with scaled branches. The tree is read from left to right; the vertical axis has no meaning. Reading from left to right corresponds to the degree of similarity, calculated here by the DICE coefficient; the scale is shown on the upper left-hand side over the branches (the horizontal lines). The vertical lines mark the degree of similarity among the branches. The first four isolates, noted by the bracket, have 100% similarity, that is, in a band by band

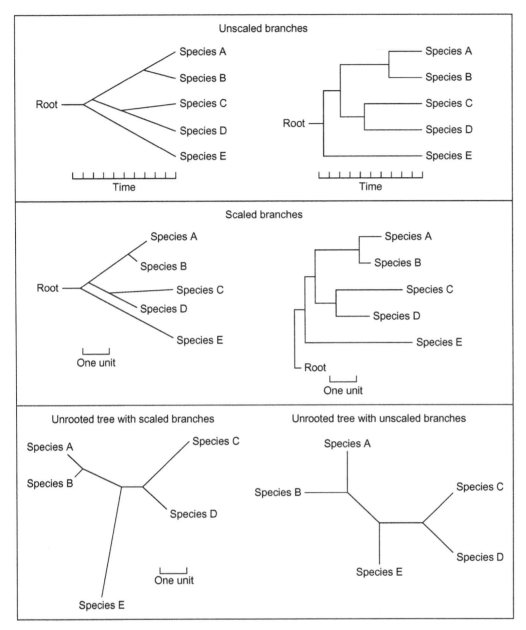

**FIGURE 7.3**
Examples of dendrograms with unscaled and scaled branches, and rooted and unrooted trees. *Source: Reproduced, with permission, from Vierstraete[16] (1999).*

comparison, the patterns are identical. These four isolates are less than 80% similar to the patient strain XMON-12, the next closest pattern.

For rapidly mutating microbes such as HIV and hepatitis C virus (HCV), order of descent is presumed to correspond to order of transmission. Phylogenetic analyses have been extremely valuable in confirming the source of HIV and other rapidly mutating viral outbreaks where there are rapid changes in viral sequence even within a single host. A molecular analysis confirmed transmission of HCV within an orthopedic ward. The HCV variants isolated from patients were closely related, and the extent of variation was consistent with the timeframe of transmission.[18] An analysis comparing HIV with simian immunodeficiency virus (SIV) suggests that HIV-1 and HIV-2 have different origins: HIV-1 is closer to the SIV found in chimpanzees and HIV-2 to that of sooty mangabey monkeys. Note that there were at least

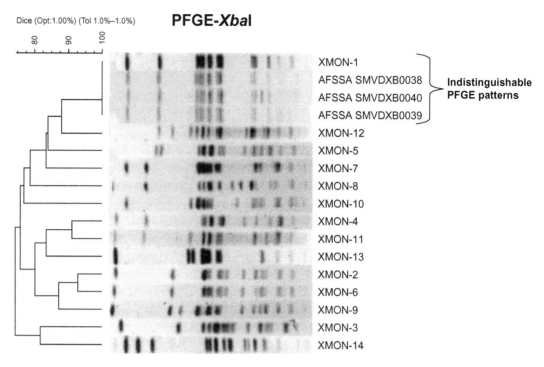

**FIGURE 7.4**

Phylogenic analysis of pulsed-field gel electrophoresis (PFGE) patterns from *Salmonella enterica* serotype Montevideo strains from an outbreak in France, 2006. Twelve case-patients and three isolates from cheese or raw milk processed in the incriminated plant (AFSSA SMVDXB0038-39-40) identified from epidemiologic analyses as the putative source shared the identical PFGE pattern (only patient strain XMON-1 is shown on the figure because all others were identical). Isolates from other case-patients (XMON-11-12-13-14) and non-outbreak-associated patients (XMON-2 to XMON-10) were quite heterogeneous. Similarity calculated using DICE. *Source: Reproduced, with permission, from Dominguez et al.[17] (2009).*

three separate introductions of each virus, and that the genetic distance even within HIV-1 of the same subtype (group O, for example), is reasonably large (Figure 7.5).

As it becomes progressively easy to obtain genetic sequences, phylogenetics is increasingly being applied in epidemiology. Inferring phylogeny and testing the likelihood that a specific model of sequence evolution is correct is quite complex. Therefore it is desirable to collaborate with an expert in the area. Once a tree is constructed, however, the shape of a tree can give insight into whether a trait derived from a common ancestor was laterally acquired or independently evolved, which is an important question for understanding the emergence and spread of a pathogen. The structure also reveals underlying epidemiologic processes that result in different transmission patterns.

Grenfell and colleagues[2] compared the phylogenetic structure of measles, dengue, HIV, and HCV (Figure 7.6). Although there is a vaccine against the measles virus that is highly effective against all known variants, it is equally effective against all variants. This is reflected in the phylogenetic structure of measles: several different variants are present at any one time (Figure 7.6A). This is in contrast to influenza A, where there is only partial immunity to strain variants. This selects for variants where there is little or no immunity. Variants become extinct at high rates, and only a few strain variants are present at any one time (Figure 7.6B). The four different serotypes of dengue are equally distant from each other in a genetic sense; there are multiple strains within each serotype that are clustered close together (Figure 7.6C). The phylogeny of HIV and HCV at the population level is very similar (Figures 7.6D and 7.6E). Individuals can be infected with more than one strain, meaning cross-immunity is limited. Clusters reflect transmission patterns rather than selection from host immunity. By contrast, the immune system of an infected host provides strong selective pressure. The within-host HIV phylogeny has a shape similar to the population phylogeny similar to that of influenza (Figure 7.6F).

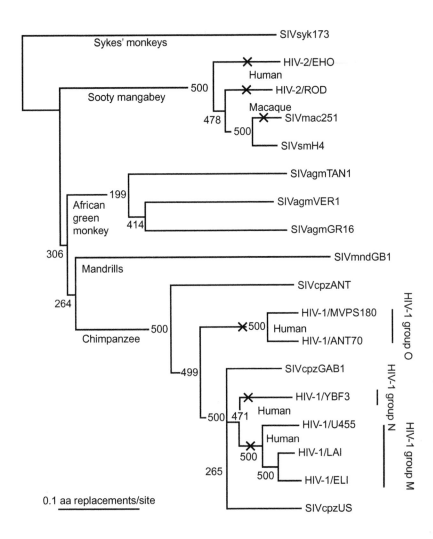

**FIGURE 7.5**
The relationship of HIV to representative simian immunodeficiency viruses using the viral polymerase gene (Pol) amino acid sequence. Crosses represent possible transmission between species. The length of the distance bar is ~1 amino acid replacement per 10 sites. Horizontal lines represent the degree of genetic distance, vertical lines have no meaning. The tree is rooted using Sykes' monkeys. *Source: Reproduced, with permission, from Hungnes et al.[14] (2000).*

Because they evolve rapidly, the most extensive epidemiologic applications of phylogenetics have been applied to viruses. By analyzing the tree structure, the origin, transmission patterns, and epidemic behavior have been discerned,[19,20] and the basic reproduction number, $R_0$, estimated. $R_0$ is the average number of secondary cases that arise from a primary case in a susceptible population. To be maintained in a population, the $R_0$ for an infection must be 1.0 or greater; if greater than 1.0 the number of infections is increasing (epidemic), if equal to 1.0 the infection is at a steady state (endemic). Estimates of $R_0$ are thus extremely useful for predicting the course of an epidemic. $R_0$ is estimated using mathematical models, most easily for infections that induce lifelong immunity using the age of first infection as a parameter. For viruses that rapidly evolve, $R_0$ can be estimated using a maximum likelihood method, which also enables estimation of the population size required for persistence.[19] The insights into origin, transmission patterns, and epidemic behavior help predict the course of ongoing epidemics and emergence of new variants.

The reader is referred to *Molecular Evolution: A Phylogenetic Approach*[21] and *Phylogenetic Trees Made Easy*[22] for an introduction to phylogenetics.

## 7.7 ANALYSES OF MICROBIOME DATA

The potential contributions of the omics to infectious disease epidemiology are not limited to studies of single organisms. Next generation sequencing technologies like the 454 pyrosequencing (Roche, Branford, CT) and massively parallel sequencing by synthesis (Illumina, San Diego, CA) make it possible to characterize the membership and structure of microbial communities based on genetic sequence. Studies use the sequence of the

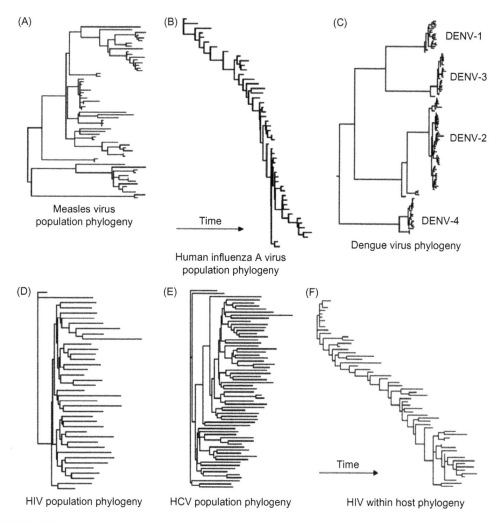

(A) Measles virus population phylogeny

(B) Time → Human influenza A virus population phylogeny

(C) DENV-1 DENV-3 DENV-2 DENV-4 Dengue virus phylogeny

(D) HIV population phylogeny

(E) HCV population phylogeny

(F) Time → HIV within host phylogeny

**FIGURE 7.6**

**A**, Measles phylogeny: the measles virus nucleocapsid gene (63 sequences, 1575 base pairs [bp]). **B**, Influenza phylogeny: the human influenza A virus (subtype H3N2) hemagglutinin (HA1) gene longitudinally sampled over a period of 32 years (50 sequences, 1080 bp). **C**, Dengue phylogeny: the dengue virus envelope gene from all four serotypes (DENV-1 to DENV-4, 120 sequences, 1485 bp). **D**, HIV-1 population phylogeny: the subtype B envelope (E) gene sampled from different patients (39 sequences, 2979 bp). **E**, HCV population phylogeny: the virus genotype 1b E1E2 gene sampled from different patients (65 sequences, 1677 bp). **F**, HIV-1 within-host phylogeny: the partial envelope (E) gene longitudinally sampled from a single patient over 5.8 years (58 sequences, 627 bp). All sequences were collected from GenBank and trees were constructed with maximum likelihood in phylogenetic analysis using parsimony (PAUP). Horizontal branch lengths are proportional to substitutions per site. *Source: Adapted, with permission, from Grenfell et al.[2] (2004).*

ribosomal genes (designated rDNA for ribosomal DNA), which are present in all cellular organisms, to characterize the communities. All cells have ribosomes, and the genes are highly conserved so they can be used for taxonomy. Rapid sequencing technologies give an unprecedented opportunity to characterize the microbial communities living on and in the human body, how they vary over short and long periods, respond to treatment, and interact with pathogens. The Human Microbiome Project is an ongoing effort funded by the National Institutes of Health to provide these parameters. The goals are:

- Determining whether individuals share a core human microbiome,
- Understanding whether changes in the human microbiome can be correlated with changes in human health,
- Developing the new technological and bioinformatic tools needed to support these goals, and

**TABLE 7.6 Ecological Terms and Parameters Used in Genetic Analysis of Microbial Communities**

| | |
|---|---|
| Operational Taxonomic Unit (OTU) | Classification that distributes individuals into groups based on some criteria |
| Phylotype | Classification based on phylogenetic relationships |
| Community | An assemblage of different species in the same place and time |
| Metacommunity | The larger community that encompasses spatially separated communities |
| Population | A group of individuals of the same species in the same place and time |
| Metapopulation | The larger population that encompasses spatially separated populations |
| Structure | A description of the OTUs in a community and their relative abundance |
| Richness | The number of different OTUs in a community |
| Evenness | A measure indicating the relative amounts of each OTU in a community |
| Diversity | A measure indicating both the richness and evenness of the community |

- Addressing the ethical, legal and social implications raised by human microbiome research (http://nihroadmap.nih.gov/hmp/).

Taxonomy based on ribosomal gene sequence does not map perfectly to taxonomy based on other methods. Bacteria were originally placed into taxonomic groupings based on phenotype. Genetic analyses have clearly demonstrated that speciation based on phenotype has only a weak relationship with groupings based on genotype. Grouping bacteria based on the sequence of the 16S rRNA gene that codes for the ribosome, or only part of this gene, results in groups that correspond only at a very general level with taxonomic groupings. Although the distribution of the resulting groups can be informative, the relationship of a group to a species may be minimal at best. Therefore the resulting groups are referred to as operational taxonomic units (OTUs), that is, groupings used in lieu of species (Table 7.6). If sequences are classified based on genetic relationships using phylogenetics, the OTUs are referred to as phylotypes.

The microbiota (also known as microbial flora) living on and in the human body contains bacteria, archaea, eukaryotes, and viruses. A microbial community refers to the mix of organisms present at a time and place, such as the microbial community living in the human nose, which can include several different bacterial and viral species. This is in contrast to a population, which refers to a single species or OTU. *Staphylococcus aureus* can be isolated from the nose, armpit, pubis, vagina, and bowel. The *S. aureus* in the nose is a population; presuming that nasal *S. aureus* is swallowed and wiped on hands and thus interacts, the different populations of *S. aureus* on a human body are a metapopulation. Similarly, the microbial communities in the nose interact with communities at other body sites; the microbiota on a human are a metacommunity.

The structure of a microbial community is a measure of the OTUs present and their relative abundance. Ecologists ask questions regarding the resistance of a community to disruption and the resilience following disruption. For epidemiologists, this translates into asking questions such as the following. (1) What is the role of nasal microbiota in preventing colonization by a potential pathogen? (2) Does existing microbiota enhance or reduce risk of pathogen transmission? (3) If the nasal microbiota is disrupted by an invading pathogen (and additionally by treatment of that pathogen), does it return to its previous state? To summarize these changes, populations are characterized with respect to richness, evenness, and diversity.

Richness refers to the total number of OTUs present. Because the number of OTUs observed depends on the number of individuals sampled, to compare species richness in different samples, data are displayed in a rarefaction curve. Rarefaction curves plot the number of OTUs observed by the number sampled. The richness can then be compared. For pyrosequencing, the number sampled would be the number of sequences read. Claesson and associates[23] analyzed the intestinal microbiota in four fecal samples using pyrosequencing

113

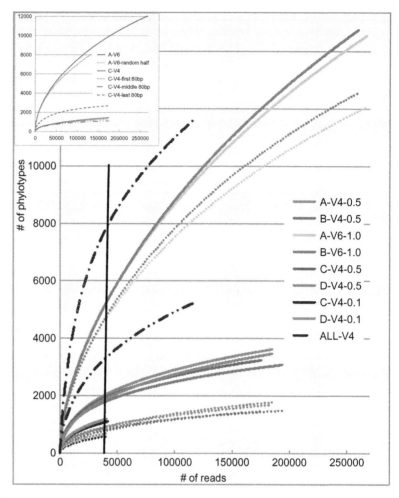

**FIGURE 7.7**
Rarefaction curves for pyrosequencing reads of the V4 and V6 regions of 16S rDNA from fecal samples from individuals A, B, C, and D. Phylotypes assigned using 97% (dotted lines) and 98% levels (solid lines, except for ALL-V4, which has single dots) of similarity. The inset shows curves for half the A-V6-1.0 reads and the three constituent parts of the C-V4-0.5 reads.
*Source: Reproduced, with permission, from Claesson et al.[23] (2009).*

of two different regions of the 16S rRNA, V4 and V6. The V6 region is known to be more variable, which therefore results in more OTUs than a less variable region. The rarefaction curve shows that for all four individuals (A, B, C, and D) at the same number of reads (horizontal line), there are more phylotypes observed for the V6 than V4 region (Figure 7.7). As a validation, the authors randomly selected half the reads and recreated the rarefaction curve (inset). This analysis confirmed that the V6 region is more variable.

Evenness refers to the distribution of each OTU within the sample. High evenness suggests there are similar numbers in each OTU and low evenness suggests the reverse. In the study by Claesson and associates,[23] the evenness varied from 0.51 to 0.70, using the same region and level of similarity for classification into phylotypes. Sample B was most even and sample D the least. Diversity is a function of richness and evenness; a frequently used index of diversity is the Shannon index. Diversity was highest in sample A and lowest in Sample D.

These technologies and analyses have applications beyond characterizing human microbiota. They can be used to assess environmental exposures, such as microbial composition in air

**TABLE 7.7** Documentation Standards and Associated Websites

Minimum Information About a Microarray Experiment (MIAME) standards
http://www.mged.org/Workgroups/MIAME/miame.html

MIBBI: Minimum Information for Biological and Biomedical Investigations
http://www.mibbi.org/index.php/Main_Page

Minimal Information about a Genome Sequence (MIGS)
http://mibbi.org/index.php/Projects/MIGS

HUPO Proteomics Standards Initiative (PSI)
http://www.psidev.info/

The Metabolomics Standards Initiative (MSI)
http://msi-workgroups.sourceforge.net/

Minimal Information About a Phylogenetic Analysis (MIAPA)
http://mibbi.org/index.php/Projects/MIAPA

Sharing Data from Epidemiologic Studies
http://acepidemiology.org/policystmts/DataSharing.pdf

Data Documentation Initiative (DDI)
http://www.disc-uk.org/docs/DDI_and_IRs.pdf

pollutants or water, or infectious content on fomites in clinical settings. They can be used to measure specific genes, such as those for antimicrobial resistance or virulence within microbial communities independent of species. Although it is hard to predict how an increased understanding of microbial community structure will change epidemiology and public health practice, it is a given that we are in for some surprises.

## 7.8 DOCUMENTING GENETIC, MOLECULAR, AND EPIDEMIOLOGIC DATA SETS

The initial high costs of genetic sequencing and microarray technology have had the happy side effect of increasing pressure for individual groups to rapidly share their raw data with others. A variety of public access databases have been developed, and U.S. databases are available on the National Center for Biotechnology Information (NCBI) website. Currently there are nucleotide, protein, structure, taxonomy, genome, expression, and chemical databases. The NCBI also includes a variety of educational resources, software, and user support. The expectation of sharing of genetic sequence, and results of high-throughput experiments such as microarrays, by depositing in public use databases, greatly accelerates scientific development. Sharing also enforces uniform standards for documentation and of associated jargon. There is an increasing trend for data sharing for all types of data sets, including those from the social sciences, health, and biomedical research including epidemiology. The development of minimum standards provides a documentation guide to new investigators, because presumably the standards represent the minimum that any one would want for his or her data.

The first standard developed was the Minimum Information About a Microarray Experiment (MIAME). This standard suggests that both raw and normalized data be made available, associated biological information (annotation) about what is on the array, the experimental design, specimens, and conditions. MIAME has spawned similar standards for documenting other types of high-throughput experiments (Table 7.7). Also included in Table 7.7 are statements regarding documentation of social science and epidemiologic studies.

# References

1. E.M. Volz, S.L. Kosakovsky Pond, M.J. Ward, et al., Phylodynamics of infectious disease epidemics, Genetics 183 (4) (2009) 1421–1430.

2. B.T. Grenfell, O.G. Pybus, J.R. Gog, et al., Unifying the epidemiological and evolutionary dynamics of pathogens, Science 303 (5656) (2004) 327–332.

3. K. Yokoyama, K. Makino, Y. Kubota, et al., Complete nucleotide sequence of the prophage VT1-Sakai carrying the Shiga toxin 1 genes of the enterohemorrhagic *Escherichia coli* O157:H7 strain derived from the Sakai outbreak, Gene 258 (1-2) (2000) 127–139.

4. P. Roumagnac, F.X. Weill, C. Dolecek, et al., Evolutionary history of *Salmonella typhi*, Science 314 (5803) (2006) 1301–1304.

5. H. Tettelin, V. Masignani, M.J. Cieslewicz, et al., Genome analysis of multiple pathogenic isolates of *Streptococcus agalactiae*: implications for the microbial "pan-genome," Proc. Natl. Acad. Sci. U.S.A. 102 (39) (2005) 13950–13955.

6. M. Touchon, C. Hoede, O. Tenaillon, et al., Organised genome dynamics in the Escherichia coli species results in highly diverse adaptive paths, PLoS Genet. 5 (1) (2009) e1000344.

7. M. Uttamchandani, J.L. Neo, B.N. Ong, et al., Applications of microarrays in pathogen detection and biodefence, Trends Biotechnol. 27 (1) (2009) 53–61.

8. S.B. Pounds, Estimation and control of multiple testing error rates for microarray studies, Brief. Bioinform. 7 (1) (2006) 25–36.

9. E.L. Brodie, T.Z. Desantis, D.C. Joyner, et al., Application of a high-density oligonucleotide microarray approach to study bacterial population dynamics during uranium reduction and reoxidation, Appl. Environ. Microbiol. 72 (9) (2006) 6288–6298.

10. A. Kistler, P.C. Avila, S. Rouskin, et al., Pan-viral screening of respiratory tract infections in adults with and without asthma reveals unexpected human coronavirus and human rhinovirus diversity, J. Infect. Dis. 196 (6) (2007) 817–825.

11. N. Berthet, I. Leclercq, A. Dublineau, et al., High-density resequencing DNA microarrays in public health emergencies, Nat. Biotechnol. 28 (1) 25–27.

12. L. Zhang, U. Srinivasan, C.F. Marrs, et al., Library on a slide for bacterial comparative genomics, BMC Microbiol. 4 (2004) 12.

13. J.A. Maynard, R. Myhre, B. Roy, Microarrays in infection and immunity, Curr. Opin. Chem. Biol. 11 (3) (2007) 306–315.

14. O. Hungnes, T.O. Jonassen, C.M. Jonassen, et al., Molecular epidemiology of viral infections. How sequence information helps us understand the evolution and dissemination of viruses, APMIS 108 (2) (2000) 81–97.

15. National Center for Biotechnology Information: Just the Facts. A Basic Introduction to the Science Underlying NCBI Resources. Systematics and Molecular Phylogenetics [website], Available at: http://www.ncbi.nlm.nih.gov/About/primer/phylo.html (Updated 1.04.04; accessed 10.08.10).

16. A. Vierstraete, Phylogenetics [website], Available at: http://users.ugent.be/~avierstr/principles/phylogeny.html (Updated 11.08.99; accessed 10.08.10).

17. M. Dominguez, N. Jourdan-Da Silva, V. Vaillant, et al., Outbreak of *Salmonella enterica* serotype Montevideo infections in France linked to consumption of cheese made from raw milk, Foodborne Pathog. Dis. 6 (1) (2009) 121–128.

18. R.S. Ross, S. Viazov, Y.E. Khudyakov, et al., Transmission of hepatitis C virus in an orthopedic hospital ward, J. Med. Virol. 81 (2) (2009) 249–257.

19. O.G. Pybus, M.A. Charleston, S. Gupta, et al., The epidemic behavior of the hepatitis C virus, Science 292 (5525) (2001) 2323–2325.

20. E.C. Holmes, Evolutionary history and phylogeography of human viruses, Annu. Rev. Microbiol. 62 (2008) 307–328.

21. R.D.M. Page, E.C. Holmes, Molecular Evolution: A Phylogenetic Approach, Blackwell-Science Ltd., Oxford, 1998.

22. B.G. Hall, Phylogenetic Trees Made Easy: A How-To Manual, third ed., Sinauer Associates, Incorporated, Sunderland, 2007.

23. M.J. Claesson, O. O'Sullivan, Q. Wang, et al., Comparative analysis of pyrosequencing and a phylogenetic microarray for exploring microbial community structures in the human distal intestine, PLoS One 4 (8) (2009) e6669.

# Determining the Reliability and Validity and Interpretation of a Measure in the Study Populations

There are many reasons for using a molecular test in an epidemiologic study. The test might be used to confirm or exclude disease, assess disease severity, or identify the precise location of disease. It might characterize how the host responds to disease or how microbes interact with each other or with the host. The test might be used to distinguish between disease types, identify microbes, or differentiate among microbes of the same species. The test might detect a known or suspected marker of disease prognosis, a known or suspected marker of exposure that modifies disease risk, or one that is known or suspected to interact with a defined exposure disease relationship. Alternatively, the measure might detect a marker in search of an association. New technologies have identified genetic, transcript, and protein variants whose association with disease is unknown. Regardless of the reason for measurement, a test is only useful if it is reliable and valid, and interpreted appropriately. The validity and reliability of a test result depend on everything from whether the specimen was collected correctly to whether the results were recorded accurately. Once a test is selected and the validity and reliability determined in the hands of the research team, the continued validity and reliability is assured by quality control and quality assurance procedures (discussed in Chapter 10).

The extent that a test result reflects the true value, that is, it is valid, depends on minimizing two major classes of error: systematic error (also known as bias) and random error (Figure 8.1). Systematic error is an error that occurs in one direction, for example, a scale that shows the weight as two grams higher than the true value. Though a systematic error can be corrected post hoc if detected, it is best avoided by regularly calibrating instruments. Depending on the direction of the bias, a systematic error can lead to the overestimation or underestimation of the frequency of exposure or disease. However the primary concern is if the systematic error occurs differentially between cases and controls or exposed and unexposed individuals. This difference can lead to an erroneous association. For example, if values are always higher for cases, it will appear that there is an association with being a case, even if the effect is due solely to systematic error in measurement. A well-designed and well-implemented study protocol will avoid systematic errors by (1) setting inclusionary and exclusionary criteria that result in unbiased selection and follow-up of study participants; (2) making collection, storage, and processing of specimens from all participants as similar as possible; and

**117**

**FIGURE 8.1**
Random and systematic errors (bias). *Source: Adapted, with permission, from Atkins[1] (2000).*

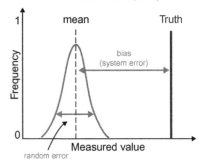

Molecular Tools and Infectious Disease Epidemiology.
© 2012 Elsevier Inc. All rights reserved.

---

**TABLE 8.1** Determining an Optimal Study Protocol

- Identify all data handling and processing steps, from specimen collection to recording data in a database
- Assess the potential for error at each step, and the error tolerance
- Determine the reliability of the selected measure across a range of values
- Determine the validity of the selected measure
- Determine the intralaboratory and interlaboratory reliability
- Determine the appropriate interpretation of measurement

---

(3) arranging laboratory procedures so that any effects of storage or testing equally impact specimens from cases and controls or exposed and unexposed participants.

The reliability of a test, that is, the extent that the same test yields the same value or very close value on repeated testing, is affected by the extent of random error. Random error increases variation and decreases the ability to detect a difference between groups – if a difference exists. Random errors can be minimized, but not avoided entirely, by minimizing technical variability in procedures, using uniform reagents and instruments, and training and periodically retraining personnel in the study protocol. Ideally, each specimen will be collected, handled, processed, and tested in exactly the same way from each study participant. If we could attain this ideal, all observed differences between participants would be attributable to true biological variation – given a test with perfect reliability and validity. After ensuring that the protocol does not introduce systematic error, our goal is to develop a study protocol and laboratory procedures that minimize random error.

This chapter describes the steps towards developing an optimal study protocol for a valid and reliable test result, and identifying any issues of interpretation of the selected measure for the study population (Table 8.1).

## 8.1 IDENTIFY ALL DATA HANDLING AND PROCESSING STEPS, FROM SPECIMEN COLLECTION TO RECORDING DATA IN A DATABASE

Identifying all data handling and processing steps is essential for establishing a study protocol that minimizes error. This listing is also required for establishing quality control and quality assurance procedures intended to minimize errors throughout the conduct of the study (see Chapter 10). The listing should be as exhaustive as possible (see Table 8.2 for an example from a study of group B *Streptococcus*).

The amount of observed variation for a specific test depends on how sensitive the test results are to all the steps up to and through laboratory testing. Variation also is affected by the type of specimen, the quality of the specimen, the test itself, and how the results are to be interpreted. Results from tests of liquid specimens can be quantitative, because the test can be referred back to a sample of known volume. The amount of bacteria isolated from urine is reported as the number of colony-forming units per milliliter urine. Tests of solid specimens may give only qualitative results. The amount of bacteria isolated from a throat swab is reported as heavy, moderate, little, or no growth. It is possible to process solid specimens or that collected on swabs by dissolving, or grinding, to test a known amount. Depending on processing, results may be reported qualitatively or semiquantitatively. Although many tests give results per volume, others give values relative to some standard. For example, we can use real-time polymerase chain reaction (PCR) to estimate the total bacterial load and then the proportion of that load due to a specific species, enabling semiquantitation even if the initial specimen is solid or a secretion collected on a swab. Even if the results are quantitative the interpretation may not be. The level of discrimination detected by the measuring tool may

Determining the Reliability and Validity and Interpretation of a Measure in the Study Populations

**TABLE 8.2** Example of Listing Data Handling and Processing Steps: Specimen Collection for Study of Group B *Streptococcus* (GBS)

| Procedure | Potential for Error |
|---|---|
| Self-collection of rectal specimen using a swab, placed into transport media | • Specimen does not contain fecal matter<br>• Swab placed in wrong vial<br>• Swab/vial not labeled or improperly labeled<br>• Specimen not collected<br>• Transport media improperly prepared or outdated |
| Transport of specimen to laboratory | • Specimen lost during transport to laboratory<br>• Labeling lost during transport<br>• Specimen heated or cooled during transport<br>• Delays or inconsistent transport time |
| Culture specimen for GBS | • Error in labeling<br>• Break in sterile technique<br>• Culture too long/short; incorrect medium, temperature, or other conditions |
| Identification of GBS | • Error in identification<br>• GBS isolate grown on plate, but not selected for further testing (identification)<br>• Incorrect recording of results<br>• Break in sterile technique |
| Storage of isolates | • Error in labeling<br>• Storage media not properly prepared<br>• Incorrect storage media<br>• Break in sterile technique<br>• Storage at incorrect temperature or change in temperature that impacts isolate integrity<br>• Storage location not recorded or recorded incorrectly |

not reflect a true biological difference. There may be a wide range of values corresponding to disease or exposure so that the actual interpretation is dichotomous (diseased/not or exposed/not) or at best categorical (not diseased, possibly diseased, probably diseased, definitely diseased).

When a specimen is collected can influence study results. Many biomarkers reflect circadian rhythms; if the daily fluctuation within an individual is as large as that between individuals it is difficult to draw any inferences. One strategy is to standardize time of specimen collection. Urine varies in concentration depending upon time of day and amount of fluid the individual consumes. If the investigator wishes to measure concentration of a specific biomarker, participants might be directed to collect a first morning void, and to either record the amounts of liquid consumed or drink a specified amount of liquid the day before, which will enable the investigator to adjust for liquid proportional to the individual's size. Alternatively, participants might be directed to collect all urine voided in a 24-hour period. The assay results can be further normalized to a metabolite that is excreted at a known rate.

How a specimen is collected can also influence study results. Bacteria grow in specific niches on the human host, so variation in specimen collection may change the probability of detection or of contamination from other sites. Bacteria grow around the urethral opening; in women, vaginal fluid containing bacteria may contaminate the urine specimen. If the investigator wishes to measure the concentration of bacteria in the bladder, avoiding contamination from the urethra or vaginal discharge, he or she may consider collecting urine directly from the bladder using a needle through the abdominal wall into the bladder. This may not be necessary. If the investigator can tolerate low levels of contamination, he or she might consider collecting a clean-catch midstream urine specimen – urine is collected after cleaning the periurethral area, and urinating a small amount to minimize urethral bacteria – and/or asking women to insert a vaginal tampon before voiding to minimize vaginal discharge.

Once a specimen is collected, it will go through a variety of processing steps, depending on the study. Each step should be tracked to minimize specimen loss and to ensure that each specimen is processed appropriately. To continue with the group B *Streptococcus* example, the study should keep a record of (1) all individuals screened for participation and reason for refusal, (2) the completion of each participant of the enrollment forms (consent form, questionnaires), (3) the collection of specimens from each participant enrolled, (4) some assessment of the quality of the specimens and if no specimen was collected, (5) a listing of all materials sent to the laboratory for processing each day, (6) a listing of all materials received at the laboratory each day, (7) the processing of each specimen upon receipt, and (8) where specimens are stored.

Tracking is best done using an electronic database; but databases have their own pitfalls. The data can be lost if not backed up regularly. Data entry is not without error; numbers can be mistyped or entered into the wrong cells. I strongly recommend using a database rather than a spreadsheet to record data, as it is particularly easy to enter data in the wrong cell or to sort only a single column of a spreadsheet, resulting in errors that are very difficult to rectify. Using a barcoding system can be extremely useful, because it eliminates the need to type in a code and associated errors. However, barcodes are not foolproof, as the code can be linked to the wrong data, or the links between the barcode and the specimen identification number can be lost. Personnel must become accustomed to scanning each tube or rack before processing. Further, if the codes cannot be imported easily into the databases associated with high-throughput equipment used to test the specimen, the value is minimal.

## 8.2 ASSESS THE POTENTIAL FOR ERROR AT EACH STEP, AND THE ERROR TOLERANCE

Each step in specimen collection and processing is subject to error. Specimens can be collected by study personnel, medical personnel working with the study, or by the participant. Study personnel are under the investigators' direct supervision; they can be trained and retrained periodically to ensure protocols are followed. Further, study personnel likely will be collecting specimens frequently. By contrast, medical personnel working with the study may be too busy to learn the study protocol well, and are likely to collect specimens from study participants relatively infrequently. Thus, making step-by-step instructions readily available is essential. Self-collected specimens depend on the extent that the study participant understands the directions; quality of self-collected specimens can be excellent if the protocol includes good training for participants and easy-to-follow instructions. These instructions should be given verbally along with written instructions. If specimens are collected on-site, the instructions might be taped to the wall of the specimen collection room. Regardless of who collects the specimens, having easy-to-follow instructions, color coding or numbering materials to match steps, and providing packets containing all materials and instructions together will reduce errors in specimen collection.

Molecular tests are increasingly sensitive in a laboratory sense, that is, able to detect exquisitely small amounts of material. The increase in sensitivity can translate to increased variability, as tests are able to detect previously undetectable variations among specimens. Whether the increased variation is attributable to technical rather than biological variation should be determined before publishing study results. There are distinct biological niches. For example, the bacteria found in the mouth vary by tooth surface, and bacteria on the skin vary by body site. Regardless of specimen collected, the goal of specimen collection should be to minimize variation due to differences in how, when, and where a specimen is collected, in order to maximize ability to detect true biological variation among study participants.

The investigator should determine both the minimum and optimal amount of a specimen required for a valid test. An optimal volume enables retesting (if required either because of test failure or for validation) and storage for future use. Specimen vials can be marked so that there is an easy visual check that sufficient volume was collected. For swabs, the investigator might institute visual checks that there is material on the swab, or include culture on nonselective media in addition to selective media to ensure adequate sample was collected. With an infectious disease process it is impossible to revisit a participant at the same disease state, so it is critical to get the specimen collected right each time. During the development of the study protocol, the investigator should look for key indicators that a specimen was appropriately collected and in sufficient amounts. These indicators can be recorded and analyzed as part of quality assurance procedures during study conduct.

Not only should the amount of specimen be sufficient, but the quality of the specimen must be appropriate. What constitutes good quality depends on how the specimen will be used. Requirements for a specimen that will be cultured are different than if the specimen will be processed and DNA extracted. For culture, there must be viable cells and minimal contamination; for DNA, there should be sufficient cells of interest present. Further, different assays are more or less tolerant of how the specimen is handled and processed. There are some DNA hybridization assays with high tolerance for presence of proteins in with the DNA, such as colony blots and in situ hybridizations. By contrast, for use in a microarray format, extensive preparation of the DNA may be necessary to remove all other materials. To avoid unnecessary costs and adverse effects on sample size associated with collecting specimens that cannot be used, the investigator should explore the tolerance of the intended assays to the intended specimen handling procedures and adjust either the assay or procedures accordingly.

All specimens must be properly labeled and stored appropriately until tested. Determining the required storage conditions and tolerance for storage is an important component of protocol development. Storage conditions also include the size of the vial and type of label. A small amount of specimen stored in a large vial may result in specimen loss due to evaporation. The label should be able to tolerate the storage conditions, because some labels fall off when a vial is frozen and thawed. A single specimen may be subject to several different assays, and the storage conditions can differ by assay, for example, one test requires storage on ice until testing and another test can be reliably conducted on a specimen stored in transport media for weeks; in this case the investigator might consider splitting the sample. Sometimes the most convenient collection method limits the ability to store the sample unprocessed. In this case it might be possible to do an initial stage of processing that makes storage less problematic.

Error tolerance is a measure of the acceptable level of error. If the error tolerance is set at 1%, it implies that a value of plus or minus 1% of the ideal is acceptable. Error always occurs; one way to minimize the impact of errors is to have inherent redundancies in the system. Invariably, some individuals will enroll in the study but be unable or refuse to give the required specimens or amount of specimen. Though there may be an optimal method of collecting a specimen, a less optimal alternative may be allowed. We conducted a study of urinary tract infection where *Escherichia coli* were isolated from vaginal specimens and urine. Vaginal specimens were self-collected using a tampon. Participants were instructed to insert a tampon before urinating. This had the added advantage of minimizing contamination of the urine specimen with vaginal secretions. Although almost all women were comfortable with this protocol some had never used a tampon and were unwilling to try. We offered self-collection of the vaginal specimen using a swab as an alternative, even though this increased the potential for contamination of the urine with vaginal secretions. Similarly, some women had emptied their bladder before meeting with our study recruiter. If unable to urinate, we offered women something to drink and a place to wait until they were able to urinate, even though these urine samples would have had less time in the bladder and lower bacterial counts than if collected at a later point. However, increasing our tolerance for reasonable departures from the enrollment study protocol enabled us to maximize specimen collection.

## 8.3 DETERMINE THE RELIABILITY OF THE SELECTED MEASURE ACROSS A RANGE OF VALUES

Reliability has two components: repeatability, when repeated testing of the same specimen under the same conditions yields the same result; and reproducibility, when repeated testing of the same specimen in different laboratories yields the same result. A highly reliable test will give very similar results on repeated tests; in a statistical sense the variance is small. There is generally a range for which results of a specific test are most reliable: the reliability may vary with the assessed value, particularly at the ranges of detection (very low or very high). Thus the investigator should conduct reliability assessments across a range of values. In a reliability assessment of dot blot hybridization, the variability, although not the interpretation, was greater for *E. coli* that had greater signal intensity upon hybridization. The increased variability was partly attributable to different copy numbers of the gene under detection.[2]

A general strategy for conducting reliability assessments is shown in Table 8.3. Note that the process is iterative: there is ongoing assessment to identify sources of technical variation followed by modification of the protocol until the desired level of reliability is achieved. Reproducibility uses a similar protocol except that a set of standard unknowns is evaluated in different laboratories using the same assay (discussed in Section 8.5).

Reliability is assessed in several ways depending on whether the measure is continuous or categorical. There is no standard for comparison; we are looking for agreement between measures. For continuous variables the metric may be the standard error, or the coefficient of variation (standard deviation/mean), or the intraclass coefficient of reliability. For categorical measures we use measures of agreement, the most common being the kappa statistic, a chance-corrected measure of agreement. Most spreadsheets and software packages enable calculation of these statistics; the formulas can be examined in the software documentation or in standard textbooks of statistics.

For proper interpretation of study results, further reliability assessment is required to determine the variability from repeated samples from the same individual, and variation among individuals. Repeated samples from the same individual will indicate if the measure varies with time of day, menstrual cycle, or consumption of food or liquids; the extent that this impacts the reliability will dictate if the protocol should stipulate timing of specimen collection. The dynamics of colonization of human body sites by microbes are essentially unknown. Currently there are few estimates in the literature of how frequently there is a change in the bacterial strains (or other colonizing microbes) that commonly colonize the human gut, mouth, vaginal cavity, and skin. Also unknown is the average duration of carriage. There are some estimates for group B *Streptococcus*; colonization is very dynamic, with an average duration of carriage of ~14 weeks among women.[3] This suggests that

---

**TABLE 8.3** Conducting Reliability Assessments

1. Select representative samples from the range of values (high, medium, low, and negative)
2. Conduct experiments on ~10 replicates of each sample
3. Compare results from replicates tested in the same and different experiments
4. Compare results from replicates tested on the same and different equipment
5. Identify sources of technical variation and modify protocol
6. Repeat 2 through 5 until desired level of reliability is achieved (~coefficient of variation <5%)

assessments of reliability must be done over a short time, and that loss over a 2-week period could as easily reflect true loss as sampling error. Without similar estimates to set the sampling intervals, it is difficult to interpret whether the tests themselves are less reliable than desirable or whether there is biological variation over the testing interval. Further, as tests become increasingly sensitive it is possible that the tests will detect differences due to the test itself: collection with a swab or lavage may inadvertently modify the biota of interest.

## 8.4 DETERMINE THE VALIDITY OF THE SELECTED MEASURE

The term validity is used in multiple ways in epidemiology. The term may apply to the study itself (internal validity), generalizations from the study to other populations (external validity), or to the characteristics of a specific measure. In this chapter, the focus is on determining the validity of a specific measure.

Ideally, the validity or accuracy of a specific measure is determined relative to a gold standard; that is, a measure that reflects the true value of what we wish to measure. Because the true value is often not known, different criteria have been developed to assess whether a measure is valid (Table 8.4). The weakest criterion is content validity; this criterion is generally invoked when developing a new measurement and appears more often in the social science literature than in assessments using laboratory techniques. The assessment of content validity is by professional judgment or consensus of the field. Assessing construct and criterion validity requires conduct of studies relative to some standard. Generally the investigator uses a case–control design, and selects specimens from individuals known to have the outcome of interest and those known not to have the outcome of interest. For construct validity the specimens will be positive or negative for some relevant characteristics of the phenomena, such as another test, and the results of the study will be the correlation or agreement between the two measures. For criterion validity, the specimens will be positive or negative for the phenomena as defined by the current standard ("truth"), and the results will predict the sensitivity, specificity, and predictive value of the new test relative to truth.

Antibiotic resistance is, at this writing, usually assessed using a phenotypic test, that is, a test assessing whether a bacterial species can grow in the presence of an antibiotic. As we move increasingly toward rapid testing using nonculture techniques like PCR, phenotypic tests become less practical because they require more time because the microbe must be grown. Alternatives, such as identifying the presence of a gene that causes resistance, can be used if the gene is known. *Streptococcus pneumoniae* resistance to penicillin is known to be caused by variations in penicillin binding proteins. A PCR-based test that assessed these variations would have content validity, because penicillin is known to bind to bacterial cell walls at the site of penicillin binding proteins. If presence of mutations in binding proteins correlates with resistance phenotype, then the test would have construct validity. Finally, if variants in penicillin binding proteins predict resistance based on a phenotypic test, the measure would have criterion validity.

**TABLE 8.4** Types of Validity and Their Assessment

| Type of Validity | Definition | Assessment |
|---|---|---|
| Content | Pertains to underlying biologic phenomena | Professional judgment, consensus |
| Construct | Correlates with relevant characteristics of the phenomenon | Correlation or agreement |
| Criterion | Predicts an aspect of the phenomenon | Sensitivity, specificity, and predictive value |

Although nonculture techniques show great promise for rapid detection of antibiotic resistance for many organisms – meeting all three types of validity – they have some disconcerting limitations. Phenotypic tests for antibiotic resistance detect resistance regardless of mechanism. A PCR-based test is limited to a specific genetic mechanism of resistance. Thus, if not all mechanisms of resistance are included in a test or a new mechanism emerges, the test will be incapable of detecting that the bacterium is resistant. Further, a mechanism may be present and not active because the gene is degenerate, there is a missing regulator, or there is some other reason. In this case the test would be falsely positive.

The extent that a measure is valid is generally estimated using sensitivity and specificity (Table 8.5). Should there be no reference standard (gold standard) the extent that the new measure agrees with the old or the correlations between the measures might be reported. Ideally, the sensitivity and specificity should approach 100%. There is a trade-off between the two, however, and the cutpoint chosen depends on the type of test, the population under study, and whether the test is for screening or diagnostic purposes.

Even a test that approaches the ideal may result in error, and the extent of that error depends on the prevalence of the item of interest in the study population. This is measured by the predictive value positive, which is interpreted as the probability that the result is truly positive given a positive test. Consider a test that is 99.9% sensitive and 99.9% specific. If the prevalence of the item in the study population is 5% the predictive value positive is 99.1%. This translates to 9.5 truly negative individuals out of every 10,000 screened being misdiagnosed as positive. The predictive value negative is even higher, 99.9%, but 0.5 truly positive individuals will be misdiagnosed as negative. For a prevalence of 1%, the predictive value positive is 91.0% and predictive value negative remains 99.9%. Should the sensitivity and specificity be only 95% each, the predictive value positive falls to 50% if the prevalence is 5%, although the predictive value negative is 99.7%. Should the prevalence be 1%, the predictive value positive drops to 16%.

For many tests there is no gold standard available. This is often the case with new tests that assay a characteristic that was previously unmeasurable, such as gene expression profiles. In this situation, after the reliability of the test is assessed, it is standard to assess the predictive validity of the test. Predictive validity is the extent that the test predicts an outcome of interest. This entails obtaining prospective specimens for testing and observing the extent the test can discriminate between those with and without the outcome of interest. Leishmaniasis is a vector-borne disease of humans and animals caused by a parasitic protozoan of the genus *Leishmania*. No gold standard is available, and there are questions regarding the validity of each of the available methods. To assess the validity of various methods, Rodriguez-Cortes and colleagues[4] used an experimental model and compared the predictive validity of each method. This assessment also pointed out the strengths and weaknesses of each test for different purposes; the best test for diagnosis was an enzyme-linked immunosorbent assay (ELISA). Quantitative PCR was useful for tracking parasite load

**TABLE 8.5** Sensitivity,[*] Specificity,[†] and Predictive Value[‡,§]

|  | True Positive | True Negative |  |
|---|---|---|---|
| *Test positive* | A | B | A + B |
| *Test negative* | C | D | C + D |
|  | A + C | B + D | A + B + C + D |

[*]Sensitivity: Probability test positive, given truly positive: A/(A + C).
[†]Specificity: Probability test negative, given truly negative: D/(B + D).
[‡]Predictive value positive: Probability truly positive, given test positive: A/(A + B).
[§]Predictive value negative: Probability truly negative, given test negative: D/(C + D).

but overall was less predictive than the ELISA test. The different tests were compared using receiver operating curves (ROCs).

## 8.4.1 Receiver Operating Curves

Receiver operating curves graphically display the trade-off between sensitivity and specificity for various cutpoints of diseased, nondiseased, exposed, nonexposed, or any test that dichotomizes a population. ROCs are used frequently in the diagnostic literature, because they enable comparison of tests and a quick evaluation of whether a test classification is better than might be achieved by chance alone. Researchers are also beginning to use them with classification rules applied to microarray data.[5] ROCs enable comparisons between tests compared to the same reference values; such comparisons might be used to either choose between tests or determine optimal test order if tests are used in series.

Any test classification has some error, because the distributions of diseased/nondiseased or exposed/nonexposed groups overlap; dichotomizing has tremendous utility but is somewhat arbitrary. The distribution of diseased and nondiseased individuals might be as shown in Figure 8.2. Setting the cutpoint at a test value of 1 erroneously places some diseased persons into the nondiseased group and vice versa.

If the threshold value was moved to a test value of −3, no cases would be missed (100% sensitive), but given the distribution of nondiseased, virtually all nondiseased would be classified as diseased (little specificity). Alternatively if the threshold was set at 3, the reverse would be the case. All positives would have disease (100% specificity) but many cases would be missed (poor sensitivity). The desired cutpoint depends on the application and what other tests are available. Using multiple tests in parallel increases the validity; a typical strategy used in diagnostic testing.

Disease stage or extent of exposure can modify the validity. For example, in persons with HIV, antibody titers are only detectable after ~3 months postinfection. Study population

125

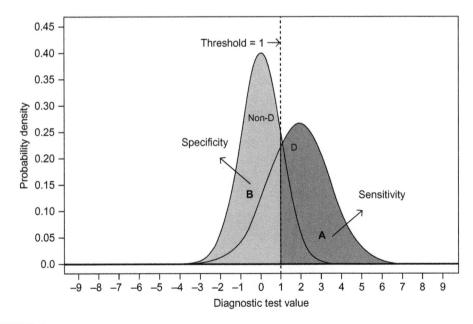

**FIGURE 8.2**

Distributions of diseased *(D)* and nondiseased individuals, shown as the probability density function, for a diagnostic test. If a test value of 1 is used to classify diseased from nondiseased, the sensitivity (the shaded area under the diseased distribution A) is 72%, and specificity (the shaded area under the nondiseased distribution B) is 84%. *Source: Reproduced, with permission, from Zou et al.[6] (2007) and Mauri et al.[7] (2005).*

variables may also modify validity so the investigator should ensure that the selected cutpoint is appropriate for the desired purpose. Some ethnic groups have different ranges of normal values than others. Tests can also be done in series, also increasing the validity as is generally done for screening. HIV screening is done in series. All individuals positive by the first test are retested with a more definitive test. Because the costs of missing a positive case are high, the sensitivity of the HIV screening test is set very high. Further tests are more specific, but less sensitive. Because the population screened using additional tests is enriched for potential cases (higher prevalence) the predictive value positive and negative will be improved. This strategy misses the false negatives, as only those screening positive are retested.

These trade-offs are visualized by plotting the sensitivity (true positive rate) versus $1 -$ specificity (the false positive rate). An ideal test hugs the $y$-axis and then moves parallel to the $x$-axis at the highest value of the $y$-axis, line A on Figure 8.3. Line A represents the gold standard. A typical test (after smoothing) looks like line B; line C is the chance line, because a test that fits that line classifies no better than chance alone. The area under the curve (AUC) gives a quantitative assessment of the accuracy of the test. If the test is perfect the AUC is 1.0; an AUC of 0.5 is considered poor. Guinto-Ocampo and colleagues[8] compared three laboratory indicators, white blood cell count, percent of lymphocytes, and absolute lymphocyte count (ALC) to a PCR test for pertussis among 141 infants who were tested for pertussis; 18 infants (13%) tested positive.[8] The ROC curves were not smoothed (Figure 8.4). Notice that ALC is closest to the upper left-hand corner, and thus predicted best of the three. The area under the curve was 81% (95% CI: 72%, 90%). The point maximizing sensitivity and specificity was an ALC cutoff of 9400; using this cutpoint the sensitivity was 89%, specificity was 75%, the positive predictive value was 44%, and the negative predictive value was 97%. Although ALC was not a strong predictor of pertussis, in the absence of a licensed PCR test for pertussis, using the ALC cutpoint provides a reasonable guide to patient management, at least while awaiting results of culture, which can take up to 10 days.

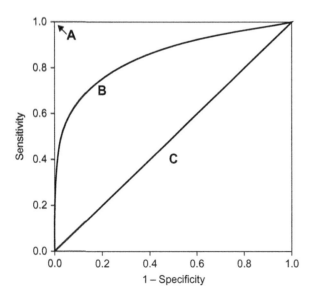

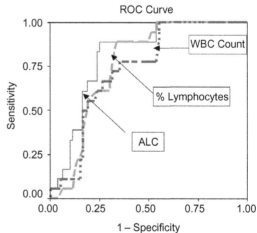

**FIGURE 8.3**

Receiver operating curves plots sensitivity (true positive rate) versus $1 -$ specificity (false negative rate). A perfect test (gold standard) is curve *A*; the area under the curve (AUC) is 100%. Curve *B* is like a typical test; the AUC is 85%. *C* is the chance line (AUC of 50%); a test on line *C* is no better than chance alone. The most accurate tests approach curve *A* and have an AUC close to 100%. *Source: Reproduced, with permission, from Zou et al.[6] (2007).*

**FIGURE 8.4**

Receiver operating curve comparing the accuracy of using white blood cell counts (WBC), percent lymphocytes, and absolute lymphocyte counts (ALC) for predicting pertussis among 141 infants tested for pertussis using a PCR test. The area under the curve for ALC was 81% (95% CI: 72%, 90%). *Source: Adapted, with permission, from Guinto-Ocampo et al.[8] (2008).*

## 8.5 DETERMINE THE INTRALABORATORY AND INTERLABORATORY RELIABILITY

*Intra-* means within and *inter-* means among or between. Random variation occurs within and between laboratories; the smallest variation is observed when replicate samples are tested in the same experiment, under identical conditions, and the largest variation is observed when samples are tested in different laboratories using different techniques (Table 8.6). There are some situations in which all specimens may be tested simultaneously in the same experiment, for example, molecular fingerprinting of bacterial isolates from a small disease outbreak. The accuracy and reliability of many rapid typing techniques, such as PCR-based techniques using random primers or repetitive elements, often depends on the ability to test isolates in a single experiment, as there can be considerable variation from experimental run to experimental run, although the findings within a run will be informative.

Intralaboratory reliability generally means assessing reliability among the technicians who will be conducting the experiments, and if there are multiple pieces of equipment, among equipment. The strategy is the same as shown in Table 8.3.

Determining interlaboratory variation is important for comparing results across geographic areas. In our global economy the potential for rapid spread of infectious agents is real, as demonstrated by severe acute respiratory syndrome (SARS). Antibiotic resistance also may emerge locally but be spread by travelers or via food transported within and between countries. Hageman and associates[9] (2003) conducted an assessment of laboratories participating in the U.S. National Nosocomial Surveillance System to validate antimicrobial testing results; 193 laboratories participated from 39 states. Laboratories were sent test organisms including an oxacillin-resistant *Staphylococcus aureus* and vancomycin-resistant *Enterococcus faecalis*; *S. aureus* and *E. faecalis* are important causes of hospital-acquired infections (nosocomial). All laboratories were able to correctly identify the *S. aureus* as oxacillin resistant; and although 88% of the laboratories identified the *E. faecalis* as resistant to vancomycin, the accuracy depended heavily on the testing method. Only 28% of laboratories using the disk diffusion methods correctly reported the result compared to 94% using minimum inhibitory concentration methods. These types of assessment studies highlight which methods work best in practice; providing feedback to participating laboratories improves diagnostic capability everywhere and the accuracy of surveillance data. Regular assessment of the validity and reliability of laboratory testing within and among participating laboratories ensures overall data quality. This topic is addressed in more detail in Chapter 10 on study conduct.

## 8.6 DETERMINE THE APPROPRIATE INTERPRETATION OF THE MEASUREMENT

The reliability and validity of a test are determined by the test developers, but these determinations are only a guide. What might be achieved in a given laboratory depends

**TABLE 8.6** Factors Influencing Observed Random Variation in Replicate Samples Tested for the Same Outcome

| Variation | Technique | Experiment | Day | Laboratory |
|-----------|-----------|------------|-----|------------|
| Least | Same | Same | Same | Same |
| | Same | Different | Same | Same |
| | Same | Different | Different | Same |
| | Different | Different | Same | Same |
| | Same | Different | Different | Different |
| | Different | Different | Different | Same |
| Most | Different | Different | Different | Different |

on the technical expertise, equipment, and population under study. It is essential that the investigator determine the reliability and validity of the test in his or her laboratory, and that these levels are monitored (via use of duplicate samples, and positive and negative controls) throughout the conduct of the study. Ideally, laboratory personnel will not know which samples are duplicates nor the exposure or disease status of the person from whom the samples were collected (a procedure known as masking or blinding). Once the investigator is confident that the test is being conducted properly, the next step is to determine the appropriate interpretation of test results. Determining the correct interpretation depends on whether the study purpose is diagnostic or exposure assessment.

Often, the intended result of a test is diagnostic – is the individual healthy or ill? As mentioned in the discussion of validity, the investigator should consider a variety of factors when interpreting the results of a particular diagnostic test, especially how the results might be modified by the characteristics of the study population. Clinical decisions are generally based on reference ranges specific to the local population, values that reflect the observed mean and 95% confidence interval. Determining reference values is standard procedure for clinical laboratories; these values are generally reported to clinicians along with test results, reflecting the local population as well as local laboratory conditions that influence test values. Multilaboratory studies thus must consider local variations in values; all participating laboratories will use the same standards (positive and negative controls), which can be used to normalize values across sites.

All tests involve some error. *Ascaris* is a parasitic round worm that lives in the intestine, consuming partially digested food. If eggs are detected in stool samples it is certain that the individual is infected. But for a variety of reasons – including the number of worms present and the distribution of eggs in the stool – parasite eggs are not always found in stool even though the parasite is present. Thus clinically, it is essential to consider the entire clinical picture, rather than the result of a single test. In some populations, presence of a single parasite ovum in the stool will not necessarily correspond to clinical disease. This does not mean that there is not some effect of parasite load on health, only that in endemic populations low levels might be considered in the normal range.

That population-specific norms are important for the clinical interpretation of a measure should be kept firmly in mind as new molecular measures are developed to characterize health and disease. Limited predictive value for a measure in one population does not preclude its ability to predict in another or across populations, perhaps indicating a causal relationship. Where malaria is endemic, a blood smear positive for plasmodia is not predictive of disease symptoms except among the very young, pregnant women, and those visiting from malaria-free areas. A cross-sectional study in an endemic population will find most individuals to have malarial parasites present. In this case, the presence of parasites in the blood might be dismissed as an incidental finding; those without symptoms also have them. Therefore it would be difficult using an endemic population to demonstrate an association between parasite presence and symptoms or even to use parasite load as a meaningful indication for therapy. When developing a diagnostic test or using a test as a potential marker of exposure or outcome in explorations of causality, the epidemiologist must consider the entire population picture.

Similar to clinical measures, many molecular measures are only appropriately interpreted within a population context. Molecular fingerprints (described in detail in Chapter 6), are an important adjunct to epidemiologic outbreak investigation. When the epidemiologic evidence is strong, the fingerprinting method is considered confirmatory, verifying that all affected individuals were infected by a microbe with the same fingerprint and it was found in the putative source (in a common source outbreak). However, in the absence of epidemiologic evidence, the presence of a common fingerprint is a much poorer predictor of true linkage, as the probability that the fingerprint might occur in the population by chance alone is much higher. This might be remedied by studies describing the distribution of fingerprints for that organism in the population for comparison. Because the generation time for microbes is much shorter than for humans, and the genomes of many pathogenic bacteria are quite heterogeneous, this would have to be an ongoing project.

Interpretation is also influenced by how close the measurement is to the construct of interest. Finding the presence of genes that code for a specific factor in an infectious agent does not guarantee that the gene is expressed in vivo. It may only be expressed under certain circumstances, the gene may not be functional because the code has been modified in ways not detectable by the test, or there must be other genes present and active for expression to occur. This is a limitation of rapid assessments for antibiotic resistance based on gene presence rather than gene expression. Tests based on gene expression also have limitations. Gene expression does not have a one-to-one correlation with protein production. Most processes require multiple steps; a test will generally focus on expression indicating one step. If other steps are not functioning, the protein of interest will not be made, or may be made but not in proper form. Even detection of the phenotype under laboratory conditions may not have a direct correlation to what is observed in vivo. Treatment of a bacterium with an antibiotic to which it is resistant in vitro may still result in clinical cure. The antibiotic may concentrate where the bacteria clusters, or the resistance mechanism is overexpressed under laboratory conditions invalidating test results, or there may be some other reason.

It is also critical to have a clear fix on what constitutes a negative test versus a failed test. A negative test implies that the characteristic of interest is not present. A failed test can look like a negative test but it fails to assess anything. The difference is enormous when interpreting study results. If a positive test is based solely on the presence of a PCR product, it is impossible to determine what the absence of the product means. Some proportion, hopefully all, are true negatives. Without proper controls, such as including primers for a genetic sequence that should always be present that gives a PCR product of a different size, it is impossible to determine if the lack of product was due to experimental error.

Proper interpretation of exposure measurement is similar to that of outcome measures. It entails an understanding of the biological variability of the measure, the inherent variability due to testing, and the construct being measured. Biological measures may vary markedly within an individual. After a night's sleep, we are all taller than after a day spent upright. If there is almost as much variation within an individual over the course of a week as observed between two individuals at a single point in time it is difficult to draw any valid conclusions.

A highly discriminatory measure makes it possible to distinguish between groups, often with great precision and statistical significance. This level of precision may be misleading, as the scale of measurement can imply a greater level of precision than is inherent in the data. The investigator is cautioned to have clear criteria for determining validity, lest an excellent test is consigned to the trash can because the measuring scale suggests it is too variable. Envision an ELISA that quantifies the amount of human chorionic gonadotropin (hCG) present in the urine. hCG is only found in the urine if a woman is pregnant. The added level of precision that quantifying the amount of hCG has no real meaning if the construct being measured is pregnancy. However, a high degree of variation may suggest a poor test, when actually it reflects stage of fetal development.

## 8.7 USING STORED MATERIALS

Assembling and following a large human cohort requires significant effort and expense, especially if biological specimens are collected. Using existing data collections and piggybacking on ongoing efforts is thus extremely attractive. However, using existing collections is not without limitations (Table 8.7). Similarly, using specimens collected for clinical testing for developing or validating new diagnostics has the advantage of

**TABLE 8.7 Considerations When Using Stored Specimens**

- Constraints imposed by study design
- Limitations imposed by study conduct (study population, selection biases)
- Reliability and validity of existing diagnosis and exposure variables
- Sample quality
- Sample quantity
- Effects of storage on substrate of interest

convenience, but there are inherent limitations that should be taken into account in data interpretation. Further, the type of specimen, length of time in storage, and method of specimen storage all influence the use of the specimens in future studies. In general, none of these are fatal flaws, but any one may substantially limit the study generalizability.

The original study design constrains what design can be used for a study of existing data. Depending on the sampling scheme, specimens from a cohort study might be analyzed as a cohort study, a cross-sectional study, or a case–control study for a variety of outcomes and exposures independent of the original study purpose. Other study designs impose stronger constraints: specimens from a case–control study are generally limited to further examining the same outcome, although the case definition might be refined after specimens are tested. If the controls were population-based or sampled from a cohort, the controls might be analyzed to give insight into the population prevalence of a variety of variables, similar to a cross-sectional study.

Different study designs impose different sampling schemes that limit the parameters that can be estimated, and the generalizability of results (see Chapter 9). Inferences made using existing specimens will have the same limitations. A study population may have been limited by age, gender, or racial/ethnic group for reasons associated with the primary study purpose. Controls may have been selected to match case characteristics; while optimizing ability to test the primary study hypotheses, specimens from controls will be highly selected and not reflect the general population. Disease risk is often modified by individual behavior; the effects of exposure may be modified or mediated by other exposures not easily ascertained from biological assays. These types of data may not have been collected and may not be easily ascertained from records.

Even a well-conducted study has limitations. Not all data will be collected for all participants for all variables and all time points. Missing data and loss to follow-up result in less than the entire sample included in analyses, often substantially reducing the effective sample size. Not only does this reduce the power to detect significant effects and the precision of estimates, patterns of loss usually are not random. The effects of such losses can bias inference and generalizability. The potential for bias can be evaluated by comparing results from analyses of the subset with complete data to a set where missing values are inputted. Selection biases occur even within repositories and data banks; reasons for participation or refusal are associated with health, access to medical care, clinical manifestations of illness, socioeconomic status, and age, which in turn are associated with disease risk.[10]

Not all variables can be accurately assessed from existing specimens. For many variables that may modify the study interpretation and generalizability of study results, the investigator will be dependent upon the quality of the data already collected. Because the purpose of the secondary analysis may be somewhat or even quite different from that of the original study, important modifying variables may not be measured with the optimum level of precision. This in turn can lead to uncontrolled confounding adversely affecting the validity of the secondary study findings.

Specimens are generally collected over time, even for a cross-sectional study, and therefore are stored for varying lengths of time. In a case–control study, controls may be identified at the same time as cases or only after the case groups is assembled. If the latter, and the assay is sensitive to this time difference, the results could suggest a difference between groups solely attributable to storage. Similarly, if specimens are sensitive to storage and are collected from individuals each year, degradation over time might be erroneously interpreted as an increase in the item of interest with time: with less time to degrade, the most recent specimens will have higher levels. These effects are difficult to assess, but some insight can be gained, if there is sufficient sample available, by retesting for measures tested before storage.

Specimens are stored in a variety of ways. They may be placed in various media or solutions and frozen, or preserved or dried and stored at room temperature. The type and extent of processing and storage all can influence the experimental results; the severity of the effect depends on whether the item of interest is DNA, RNA, protein, or a microbe that the investigator plans to regrow. DNA, RNA, or protein may alter or degrade with time. Microbes may die, or when regrown following storage display different phenotypes. Any changes in results attributable to storage are problematic because they add an additional source of variation beyond that due to technical and biological variation. Sample quality depends on how the original sample was collected, the processing it was subjected to, and the storage conditions – including times thawed and refrozen. Different laboratory tests are more or less sensitive to these factors; when using existing specimens the investigator must judge the potential impact on any inferences made. Thus it is critical to assess the validity and reliability of the planned assays on the specimens. As stored specimens are precious and often irreplaceable, these studies must be planned carefully to maximize the assessment while minimizing any losses.

Laboratory procedures might be optimized using freshly collected specimens subjected to the same handling and processing as those from the repository. Although the same length of storage probably cannot be duplicated, specimens might also be frozen and thawed, for example, to assess any effects on results relative to fresh specimens. Initial assessments of repository specimens might be made using material left over from another use, or on samples from which multiple replicates are available. These assessments can detect if the specimen contains material that inhibits or modifies the reactions in the planned experimental procedures, and, if present, to identify additional processing steps that might minimize these results. It also can determine if the specimens are of sufficient quality to be reliably and accurately assessed using the planned testing procedures.

Most tests require a minimal amount of sample for testing. If there is limited amount of sample available that is not renewable, or if there are concerns about affecting specimen integrity if thawed and refrozen, it may be difficult to obtain access to stored samples. Bacterial and viral collections are generally renewable resources, and human DNA can be immortalized, making it effectively renewable. But other microbes and biological specimens (blood, cells, tissue) generally are not renewable. The investigator will want to optimize use of nonrenewable resources, grouping secondary tests together, which may limit the amount of sample available. Even if collections are renewable, investigators are appropriately concerned that other users be aware of the collections strengths and weaknesses, and appropriately take into account any design limitations. Ensuring proper use and data interpretation takes time and effort even if data are well documented. Moreover, the group collecting the data quite rightly would like to have first crack at using it. Finally, identifying, packing, and shipping specimens to another site for testing is not without cost. Thus, it is not surprising that repositories tend to closely guard access to their collections, often requiring potential users to complete some sort of application process, and tending to favor those with whom they have some sort of social connection. Unfortunately, this can mean that specimens are not used to optimum effect, because the zeal to protect the collection can lead to little or no use.

## References

1. N.T. Atkins, More on Bias (systematic) and Random Errors [website]. Lyndonville: Lyndon State College. Available at: http://apollo.lsc.vsc.edu/classes/remote/lecture_notes/measurements/bias_random_errors.html (Updated 7.1.2000; accessed 7.1.10).

2. L. Zhang, B.W. Gillespie, C.F. Marrs, et al., Optimization of a fluorescent-based phosphor imaging dot blot DNA hybridization assay to assess *E. coli* virulence gene profiles, J. Microbiol. Methods 44 (3) (2001) 225–233.

3. B. Foxman, B. Gillespie, S.D. Manning, et al., Incidence and duration of group B *Streptococcus* by serotype among male and female college students living in a single dormitory, Am. J. Epidemiol. 163 (6) (2006) 544–551.

4. A. Rodriguez-Cortes, A. Ojeda, O. Francino, et al., *Leishmania* infection: laboratory diagnosing in the absence of a gold standard, Am. J. Trop. Med. Hyg. 82 (2) (2010) 251–256.

5. S.G. Baker, B.S. Kramer, Identifying genes that contribute most to good classification in microarrays, BMC Bioinformatics 7 (2006) 407.

6. K.H. Zou, A.J. O'Malley, L. Mauri, Receiver-operating characteristic analysis for evaluating diagnostic tests and predictive models, Circulation 115 (5) (2007) 654–657.

7. L. Mauri, E.J. Orav, A.J. O'Malley, et al., Relationship of late loss in lumen diameter to coronary restenosis in sirolimus-eluting stents, Circulation 111 (3) (2005) 321–327.

8. H. Guinto-Ocampo, J.E. Bennett, M.W. Attia, Predicting pertussis in infants, Pediatr. Emerg. Care 24 (1) (2008) 16–20.

9. J.C. Hageman, S.K. Fridkin, J.M. Mohammed, et al., Antimicrobial proficiency testing of National Nosocomial Infections Surveillance System hospital laboratories, Infect. Control and Hosp. Epidemiol. 24 (5) (2003) 356–361.

10. M. Porta, I. Hernandez-Aguado, B. Lumbreras, et al., Omics research, monetization of intellectual property and fragmentation of knowledge: can clinical epidemiology strengthen integrative research? J. Clin. Epidemiol. 60 (12) (2007) 1220–1225 discussion 1226–8.

# CHAPTER 9

# Designing and Implementing a Molecular Epidemiologic Study

Designing an epidemiologic study to address a specific research question is one of the most creative and challenging aspects of epidemiology. Theoretically, any research question might be addressed using any study design, but in practice we choose a study design based on external constraints. External constraints include access to a study population, frequency of the outcome of interest, frequency of the exposure(s) of interest, time, and budget available. A study usually is not conducted if the answer is already known, so when designing a study the investigator generally makes a series of educated guesses, balancing the need for more precise information against cost. Guesses are informed by the literature, by analogy with similar conditions, and by results from pilot studies.

Molecular epidemiologic studies are further constrained by the need to collect specimens, by the conditions imposed by the specimens to be collected and the laboratory assay of interest, the required handling and storage for the intended laboratory assays, and the associated increases in cost. Molecular tools require validation and assessments of reliability in the hands of the research team, and ongoing quality control and quality assurance procedures. The addition of molecular tools also removes some constraints: molecular testing of already collected specimens from ongoing or completed studies saves the time and expense of mounting a new study, and enables hypothesis generation and testing, and validation of laboratory assays. The trade-offs required to balance the external constraints imposed by adding molecular tools while maximizing ability to accurately estimate the parameters of interest is the focus of this chapter. The chapter begins with some general comments on planning a study, then examines in detail the trade-offs required when designing and implementing a molecular epidemiologic study. For a review of epidemiologic study designs, the reader is referred to Chapter 4. The chapter closes with a discussion of choosing molecular fingerprinting techniques for discriminating among microbial strains.

## 9.1 OPERATIONALIZING A RESEARCH QUESTION

A good study begins with a well-defined research question. Good research questions move science forward, fill holes in the literature, and can be feasibly studied using current methods while attending to ethical standards. Identifying the research question, however, is just the beginning. The next step entails operationalizing the question, which means identifying the parameters and study design that will answer the question posed. One strategy for doing so is to answer, in order, the questions shown in Table 9.1.

The first question requires you to think ahead to the end of the study. When you complete the study, and are holding a press conference to announce your results, what do you want to

Molecular Tools and Infectious Disease Epidemiology.
© 2012 Elsevier Inc. All rights reserved.

133

---

**TABLE 9.1** Operationalizing Your Research Question

- What inferences would you like to make when the study is over?
- What parameters do you need to estimate to make the desired inferences?
- What study design will allow you to estimate those parameters?

---

be able to say? That the frequency of a particular outcome or exposure is high in the study population? The rate of outcome in the population? That there is an association between an exposure and the prevalence of outcome? That the exposure causes the outcome? Once the inference is decided, the parameters to be estimated become clear. For example, to speak to the frequency or distribution of an outcome or exposure in a population requires estimating the prevalence, or the mean, range, or other characteristics of the measure. The parameters to be estimated largely determine the study design (see Table 4.2 in Chapter 4 for a summary of parameters that can be estimated by study design). The parameters also determine, in part, the measuring instrument. There are multiple methods for measuring almost every exposure and outcome. Most measures are modified by other variables, so choice of measuring instrument also dictates what *other* measures must be included.

## 9.2 STUDY DESIGN TRADE-OFFS ASSOCIATED WITH INCLUDING MOLECULAR TOOLS

Designing a study involves more than deciding if the study will take the form of an experiment, cohort, case–control, cross-sectional, or ecological study. It involves making a series of choices about the study population, sampling, measuring instruments, and study protocol. Each of these decisions affects the final results. And because many of these decisions affect the others, the study design process is an iterative one. The investigator must obtain access to the desired study population or samples and design a set of protocols that will work in the selected study population and meet ethical standards. The study population must be sufficiently large so that the desired number of participants can be obtained, and there must be feasible ways to identify potential participants. Access to multiple hospitals or clinics may be necessary to enroll a sufficient number of participants, which increases costs. Data collection at multiple sites adds a complexity to the conduct and analysis, because between-site variation must be assessed and then taken into account when estimating sample size and during the analysis. Compliance with protocols or follow-up varies by population. The group most likely to comply may have the lowest or highest levels of the disease of interest. The group with the highest level of the exposure of interest may have other factors that modify risk of the outcome. The presence of strong modifying factors requires either restriction in the design phase or enrolling a sufficient sample size to assess the effect of the modifying factor during the analytical phase. Restricting enrollment limits generalizability. Ensuring there is sufficient sample in each level of the modifying factor increases costs of data collection and the complexity of data analysis. This is just one trade-off. For each study design choice, the investigator balances the ideal with the feasible, altering the proposed study until an acceptable balance is achieved. Decisions may be revisited multiple times since each design choice must be balanced against the others.

The inclusion of a molecular tool requires further trade-offs that impact choice of study population and setting for data collection (Table 9.2). Molecular tools also greatly influence the study protocol, as detailed in the next section. When considering the use of molecular tools, the investigator must trade off the impact of specimen collection against the added value of the measurements that can be obtained from testing of the specimens. A molecular tool may not be required or even be the optimal measure. If the molecular markers of exposure are transient and the desired parameter is lifetime exposure, a review of medical records or self-report may be more valid. The investigator should also consider the state of knowledge; there is little value to using the latest tests if they are not validated. While it is

| TABLE 9.2 Study Design Trade-Offs Imposed by Molecular Measures |
| --- |
| • Study population and type, amount, and number of specimens collected |
| • Specimen collection and participation rates |
| • Assessment of all desired measures and respondent burden |
| • Follow-up over time versus selection bias due to loss of follow-up |
| • Sample size and specimen testing |

reasonable to consider the future value of any specimens that might be tested, the collection and storage of specimens is not without cost.

Once it is decided that specimens will be collected, it is important to decide the major assays of interest. Many molecular measures constrain how and when specimens should be collected, stored, and processed, which impacts choice of study population and setting for participant enrollment. The collection of certain specimens is more acceptable in some populations than others. The amount of blood that can be acceptably drawn from children is smaller than adults. Taking a biopsy is acceptable to someone with a medical problem, but much less acceptable to a healthy person. It is easier to time when a specimen is collected if the study participant is confined to a hospital bed than if working around a healthy person's work schedule. If a specimen must be processed immediately, there must be an acceptable field procedure to do so, or the participant must be able to travel to the study laboratory. Travel negatively affects compliance, as do any additional steps required for participation; even walking down the hall from the study recruiter to a laboratory increases the probability that the protocol will not be successfully completed.

Requiring specimen collection generally decreases participation rates. The amount that participation is decreased depends on the specimen and the setting, but a blood draw can decrease participation by 10%.[1] Increasing the number of specimens also can negatively affect enrollment, by increasing the length and complexity of the study protocol and increasing respondent burden. In addition, specimen collection can reduce compliance with follow-up, increasing potential for a selection bias to occur. There are ethical concerns as well; if the specimen will be stored for future testing of unknown types, the consent document must include this possibility. If the participant refuses to sign a blanket waiver, the investigator must ensure that specimens are disposed of following the tests that the participant did consent to, increasing costs.

A great deal of information can be gleaned from only one specimen. Human DNA may be immortalized, so that large amounts of DNA can be obtained for future use. Similarly, if a microbe is obtained from the specimen, it can be grown and tested in multiple future studies. Polymerase chain reaction (PCR) can be used to amplify DNA or RNA, which can be saved for future use. Each of these procedures, however, does introduce some potential bias as over time some changes may be introduced. Even in the absence of these processes, if there is sufficient amount of specimen it may be tested with multiple assays. However, each assay has a cost. The investigator often faces trading off the collection of multiple measures on a single individual versus fewer measures on many individuals. The associated decrease in sample size adversely affects the power of the study to detect an effect.

Finally, the collection of specimens and their handling, processing, and storage, either in the short- or long-term, greatly increases costs. The costs are felt in each aspect of the study, from increasing the value of any incentives given to participants for participation to data analysis. In the past, the highest cost of conducting a study was of participant recruitment, but some molecular tests have costs that approach recruitment, especially once the costs of analysis of the measure itself are taken into account. The benefits gleaned from these costs must be balanced against the opportunity costs. A usual trade-off is to reduce sample size, which decreases the study power to detect an effect, and limits the analysis.

## 9.3 CONSTRAINTS ON THE STUDY PROTOCOL IMPOSED BY MOLECULAR TESTING

Once the basics of the study design are in place, the design must be implemented. This requires the development of a study protocol. The protocol operationalizes the study design, detailing how the design decisions will be implemented. Study protocols outline inclusion and exclusion criteria, procedures for obtaining informed consent, tracking, and other procedures. Protocols are discussed in greater detail in Chapter 10. Our focus here is on how protocol decisions regarding when and how specimens will be collected and transported, details of storage and handling, and the intended laboratory assays are constrained by the requirements of the intended molecular tests (Table 9.3).

Ideally all eligible participants, regardless of exposure or disease status, will be equally likely to complete the study protocol. If participation or protocol completion is differentially associated with disease and exposure, the resulting parameter estimates may be biased. It is easy to imagine a case–control study where controls are less likely to give blood specimens, and controls that do give blood are more likely to have the exposure of interest. This is one argument for using hospital controls or selecting controls from those already using the health system for a case–control study, so the potential to give blood specimens is the same by group. There are other biases inherent in using hospital controls, since they are ill and likely to have risk factors that are associated with ill health (those in a hospital have higher rates of cigarette smoking than the general population), so the investigator should evaluate the trade-off between potential differential collection of specimens against other concerns. If a study population is recruited at more than one site, the implementation of the study protocol usually must be modified somewhat to meet local requirements. This can also result in different response rates by site or slight differences in the protocol for specimen collection or handling; the investigator should put protocols in place to detect if such differences affect study results (discussed in Chapter 10).

### 9.3.1 How and When a Specimen is Collected

Many tests require specimen collection using certain materials or sterile technique, or storage in a specific fashion before processing. Specimens may need to be put immediately on ice, put into media, protected from drying, processed for storage, or tested within a tight time frame. The specimens may need to be handled using sterile technique to avoid contamination. Most critically, the specimen should be collected in a way that the analyte of interest can be detected if it is present. Different handling is required if the intention is to grow and detect a microbe rather than detect its presence using PCR or to detect the transcription of specific genes.

If it is important to quantitate the amount of the analyte present in the sample, there may be additional constraints. To get a quantitative estimate of the analyte of interest, we must measure a known volume. Therefore quantitative measures are limited to liquids or solids. Liquids have the advantage that they can be mixed before measurement, increasing the probability of including the analyte of interest in the tested sample if it is present. The

**TABLE 9.3** Constraints on the Study Protocol Imposed by Molecular Testing

- How and when a specimen must be collected
- Timing of specimen collection relative to process of interest
- Specimen storage and handling
- Use of specimen for multiple tests
- Batching specimens versus testing as specimens are collected

analyte may not be evenly distributed in a solid; depending on the test, the solid might be sampled from several sites or even mixed with a liquid and a sample obtained for testing that represents the whole.

It is difficult to get a quantitative assessment from a specimen collected from swabbing a body surface, such as the throat. There are many variables in the amount sampled, including how much mucus and saliva is present, when the individual last ate, and the amount of surface swabbed. Semiquantitation, however, is possible; a swab may be used to inoculate media of known volume and a sample of known volume tested. For some body sites it is possible to collect the specimen by lavage; for example, in vaginal lavage, a known volume of liquid is introduced into the vaginal cavity and then collected. This enables a semiquantitative assessment of material present on the surface of the vaginal cavity. Nonculture techniques also enable semiquantitation relative to some standard: for example, real-time PCR gives the amounts in the sample, which can be reported relative to the amount detected of a universal primer, for example, the proportion of total bacterial load (detected by a universal primer) that is of a specific genus and species.

## 9.3.2 Timing of Specimen Collection Relative to Process of Interest

The timing of specimen collection may influence the results. Some tests are modified by the participant's biorhythms; an analyte may be highest or lowest in the morning, or vary by when the participant last ate food. Some disease indicators are detectable only after disease is established; for example, it can be several months after infection before antibodies are detected to HIV. Other disease indicators might occur before symptoms; shedding of measles virus is highest before symptoms occur. If studying disease etiology, specimens should be collected during the etiologically relevant period, that is, during the time period where disease pathogenesis occurs. These requirements put further constraints on the investigator and increase the probability that specimens will not be collected according to protocol, or will not be collected at all. The more severe the collection requirements, the greater the difficulty in collecting specimens according to protocol and in collecting all desired specimens from all participants. The investigator should consider the trade-offs in costs and loss to follow-up of trying to achieve optimal timing relative to the potential loss of information from arranging the timing of collection to be most convenient to study participants.

## 9.3.3 Specimen Storage and Handling

Once a specimen is collected the job is not over. Many laboratory tests require preliminary processing before testing; others are sensitive to how specimens are stored. There are blood tests that can only be conducted on serum, others that require blood that has been treated with heparin to prevent clotting, and others that are conducted on the cells. Further, some results change with storage conditions: some tests can only be conducted on fresh blood, or blood that has been stored at $-80\,°$F. If specimens, for example, throat swabs, will be tested for microbes, handling might range from storage at room temperature in transport media to immediate placement in skim milk followed by freezing. Some specimens require immediate processing to an intermediate point before storing, such as separation of serum from blood. Storage can result in phenotypic or genotypic changes of some microbes; if so, testing for these changes should be conducted before storage. Thus it is essential that specimen storage and handling be considered as the study protocol is developed and implemented.

Specimen handling and storage includes tracking and labeling. Logging when specimens are collected, shipped, received, stored, and tested minimizes loss (see Chapter 10, section on quality control/quality assurance). Logs can also be analyzed to identify breakdowns in procedures, for example, if the length of time from packing at one site until receiving at another exceeds ideal ranges, or if the dry ice was intact upon receipt. Labeling of original specimens and labels that enable linking back of specimen products to the original are essential.

### 9.3.4 Use of Specimen for Multiple Tests

Maximizing the potential use of each specimen is optimal. The process of collecting and handling a specimen is resource intensive, and collection of each sample potentially negatively influence study participation. However, to use the same specimen for multiple tests requires a fair amount of up-front planning to implement. Each test may have different processing and storage requirements, which requires splitting the sample. Further, it is prudent to save some sample for retesting and storage for future studies. Storage may be done immediately or following one or more processing steps, depending on what testing is foreseen in the future. Liquid samples, such as urine or blood, can be split for different tests. Because tests of blood may require the serum only, cells and serum, or storage on ice or other conditions, multiple tubes often must be collected. Fortunately this should require only one needlestick. If the specimen is collected via a swab, the swab might be placed in liquid media and the media split for testing; if a different storage is required, a dual swab which has two swabs wrapped together that function as one swab for collection can then be separated into two swabs for processing. Creating a flow diagram of processing, storage, and testing requirements for each test is a useful strategy for figuring out an optimal testing strategy (see Figure 9.1, for an

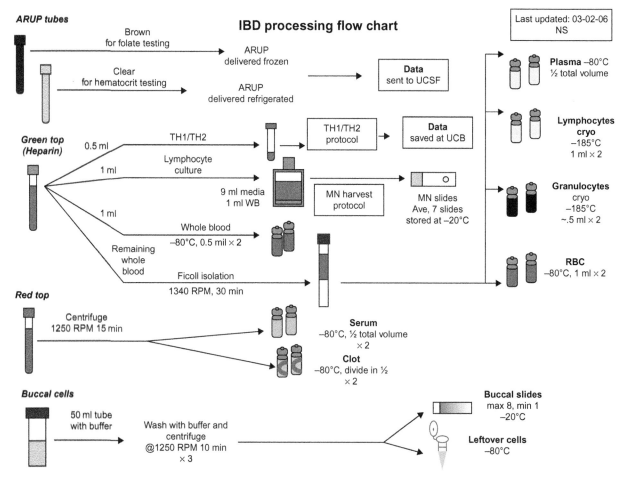

**FIGURE 9.1**

Example of specimen processing flow chart. *Source: Reproduced, with permission, from Holland et al.*[2] *(2005).*

example of a flow diagram). It is also an excellent way to identify need for quality assurance procedures (see Chapter 10).

### 9.3.5 Batching Specimens Versus Testing as Specimens are Collected

There are many excellent reasons for testing all specimens together (batching) after a study is completed. Batching enables testing within a relatively short period, so that personnel, reagents, and testing conditions should be similar for all specimens, limiting the impact of assay drift (see Chapter 10). Samples can be randomized for testing, so that the effects of the testing procedures will be distributed evenly over the entire collection, and especially across cases and controls or exposed and unexposed. But batching is not always possible; some tests must be implemented immediately, and others are sensitive to length of storage. Moreover, storage space may be limited. If data are collected over a long time period, storage requirements may make it necessary to test specimens at regular intervals. Saving testing to the end of the study also precludes interim analyses and reports, which may provide important early correction to study protocols and may delay the final report. Whether the benefits of batch testing outweigh the drawbacks depends on the constraints imposed by the specimen and the molecular test.

## 9.4 CHOOSING MEASURES OF EXPOSURE AND OUTCOME

Once it is determined that a molecular measure is desirable, the investigator must consider which measure to use. Each measure will have strengths and limitations; vary in validity and reliability; have different requirements for specimen collection, storage, processing, and technical expertise; and incur different costs. In addition to considering these factors, the investigator should consider the state of the art relative to the length of the study. If tests are under rapid development, it might be best to store specimens for testing later. It can be difficult to publish a study using a test that is no longer in favor. If this is not possible, say the specimen must be tested immediately, the investigator should select the most valid and reliable test that has the highest probability of still being in use throughout the study period. Changing testing procedures midway requires additional testing to determine the comparability of the results obtained using different measures. While this may be unavoidable, say if a manufacturer ceases production of a test or releases a new and improved version, it is best avoided whenever possible.

There are some study protocol trade-offs specific to exposure measurements (Table 9.4). Exposure in this case refers to independent variables that might be associated with or predict outcome. An exposure in one study may be an outcome in another. The trade-offs relate to whether the measure assesses current or past exposure, and the timing of specimen collection relative to the outcome of interest. The other trade-offs, regarding using a qualitative versus quantitative measure, a direct or indirect measure, and the need to access modifiers of the measure also hold for measures of outcome. The choice of measuring current or past exposure depends upon what specimen is possible to test and the exposure of interest. Exposure to some drugs can only be detected at the time of use or shortly after. Thus samples must be collected prospectively or archived samples must be available. Past exposure to a specific pathogen can be detected by testing for antibodies, and some insight into timing

139

---

**TABLE 9.4** Trade-Offs in Choosing Measures of Exposure

- Current versus ever exposed
- Timing of assessment vis-à-vis the critical period
- Qualitative versus quantitative measure
- Direct versus indirect measures
- Additional measures to assess modifiers of exposure assessment

**TABLE 9.5** Trade-Offs in Choosing Measures of Outcome

- Prevalent versus incident cases
- Time of outcome or interim measure on pathway to outcome
- Qualitative versus quantitative measure
- Direct versus indirect measures
- Additional measures to assess modifiers of outcome assessment

of infection can be gained by comparing types of antibodies. This enables detection of exposure and outcome simultaneously. If it is desired to directly identify the pathogen, it may be possible only at time of infection for a pathogen with a short duration of infection like influenza. In other cases, the pathogen remains as a commensal following infection, or is present without ever having caused disease. This is true for some *Escherichia coli* that cause diarrhea. Because of this, the presence of a diarrheagenic *E. coli* is not a good indicator of past disease. For pathogens that cause chronic infection, pathogen presence is not a good indicator of the time of infection; however, antibody titers may give some insight. The timing of specimen collection should be consistent with the understanding of when the exposure of interest acted in the pathway to disease and what can be detected in the specimen.

For measures of exposure and outcome, it is possible to use qualitative or quantitative measures. In general, there is more power when a quantitative measure is used. Quantitative measures enable detection of dose response. However, there may be additional costs associated with quantitation in terms of processing and testing. Further, the variation for some quantitative measures of biological processes may be sufficiently large that the measure is simply dichotomized in the analysis. If this is the case, it may not be worth the added costs of using a more quantitative measure. Similarly, measures of exposure and outcome can be direct or indirect. An indirect measure tends to result in more random variation than a direct measure, as it is further away from the desired measure of interest. However, an indirect measure may be much more acceptable: markers of exposure or indicators of disease that can be detected in blood or urine have great advantages over samples obtained via biopsy. Lastly, there may be variables that are known to modify the measure of exposure or outcome. These also must be measured in order to properly interpret study results. The types of modifying variables depend on the exposure or outcome of interest. A good understanding of the mechanism that leads to the specific measure is very helpful in identifying what modifying variables should be assessed.

Specific trade-offs in choosing measures of outcome parallel those specific to measures of exposure (Table 9.5). A major trade-off is whether to measure prevalent or incident cases. This is somewhat analogous to the decision of whether to measure current or past exposure. Prevalent cases include those of varying duration. For studies of etiology, using prevalent cases can be problematic as it is difficult to distinguish between risk factors for disease and risk factors for longer duration of disease. If the goal is estimating disease burden, prevalent cases are most appropriate. Molecular tools enable detection of early indicators of disease. This has the desirable effect of decreasing the length of time required for follow-up studies. However, early indicators of disease usually suggest an increased probability of the disease rather than a one-to-one correlation with disease development. Even if highly predictive, when disease will occur may be unknown, which changes the interpretation of the resulting parameter and adds an additional source of error that can reduce the power to detect an effect.

As an example of the types of trade-offs that occur when planning a protocol using a molecular measure, consider a study to determine if there is an association between group B *Streptococcus* (GBS) in the vaginal cavity and vaginal symptoms (assuming

**TABLE 9.6** Example of Trade-Offs: A Study of the Association of Group B *Streptococcus* with Vaginal Symptoms

| Research Question | Molecular Measurement |
| --- | --- |
| Detection of GBS in the vaginal cavity is associated with presence of selected vaginal symptoms. | • Culture<br>• PCR |
| The number of colony-forming units of GBS detected in the vaginal cavity is positively associated with presence of selected vaginal symptoms. | • Culture from liquid media where vaginal swab was placed |
| The proportion of GBS in the total bacterial load in the vaginal cavity is positively correlated with presence of selected vaginal symptoms. | • RT-PCR |
| Women with GBS serotype III in the vaginal cavity are more likely to exhibit selected vaginal symptoms than women carrying other GBS serotypes in the vaginal cavity. | • Culture followed by serotype detection (using either phenotypic or genotypic test)<br>• PCR |
| Selected GBS virulence factors found among GBS isolated from the vaginal cavity are positively associated with selected vaginal symptoms. | • Culture followed by either phenotypic or genotypic testing for virulence factors |

GBS, group B Streptococcus.

perfect measurement of symptoms) (Table 9.6). GBS detection can be qualitative (yes/no), semiquantitative (colony-forming units from the expressed vaginal swab), or quantitative (colony-forming units from a vaginal lavage of known volume), or assessed as a proportion of total bacterial load. The investigator might have hypotheses regarding the presence of specific phenotypic or genotypic traits on the association. Though each of these studies start with collection of a vaginal specimen, the type of specimen, how it is processed, the molecular measures used, and the ultimate interpretation of the study results will differ.

## 9.5 USING A COMMERCIAL KIT

Increasingly, commercial kits are available to detect the presence of many of the exposure and outcome measures of interest. Using a commercially available kit avoids the time and effort required to develop and validate a customized process that measures exactly the characteristic of interest. The manufacturer guarantees the quality of all reagents used in the kit, and has assessed the validity, reliability, and interpretation for some population. While this does not negate the investigator's need to determine the validity and reliability in the hands of the study laboratory, and the range of values and interpretation for the study population of interest, it does greatly decrease the amount of work required to publish test results in any reputable journal. Further, the manufacturer usually supplies technical support.

There are some disadvantages of using a commercial product. The kit may not measure the component of interest, but something related, which affects any inferences made from the study. Using kits may be more costly then purchasing reagents in bulk, and the convenience of the kit may not outweigh the cost savings. Unlike customized tests, the investigator has no control over the production of the kit. Components can be changed on short notice, or the manufacturing may be delayed or discontinued altogether. In an epidemiologic study it is essential that all measures be comparable; if a kit is discontinued, testing may need to be redone on all samples, or, at a minimum validation studies be conducted so the investigator is aware of how the test results compare between the kits. Either option can be time-consuming and impose unexpected costs.

In this day and age of mass production, almost all laboratories use commercial products. It is more economical for a large study to buy pre-prepared agar for standard bacterial cultures, for example, than to make its own. There can be subtle differences in products from different manufacturers, so it is prudent to stick to the same manufacturer for a particular product throughout the course of the study. The investigator should be alert to improvements by the manufacturer, and if a new version of the product is announced, contact the manufacturer about purchasing sufficient product of the old style to complete the epidemiologic study. There also can be variations among laboratory equipment, even those purchased from the same manufacturer. The degree of variability may be tolerable for some purposes but not for others. For example, thermocyclers – the machines used to carry out PCR – have subtle differences. For clinical tests where the output is a band that is present or absent, any differences will not be apparent in study results. However, if used for molecular fingerprinting these differences may become quite apparent and potentially bias the results, if, for example, one machine is used for cases and the other for controls.

## 9.6 MOLECULAR FINGERPRINTING

One of the most common applications of molecular tools in epidemiologic studies of infectious diseases is molecular typing or fingerprinting of the infectious agent. Molecular typing enables the investigator to discriminate among microbial strains from a single species, identify outbreaks and transmission events, and confirm the reservoir of infection. Molecular typing is most widely applied to bacterial infections, for example, in the Centers for Disease Control and Prevention PulseNet, although viruses and eukaryotic microbes also are typed using molecular tools. Key characteristics of a typing technique to consider, beyond standard reliability, validity, and cost considerations, are (1) the ability to discriminate among strains and (2) a biological basis for grouping strains with apparently different types. The level of discrimination required and the need to be able to group strains depends on the research question. There are inherent advantages in using a typing system already accepted by investigators in a particular field; it makes it possible to compare results among investigators and saves the investigator the time and resources spent developing and validating a new method. Thus any choice of typing method should also entail a thorough search of the current literature.

Discrimination refers to the number of different groups that result after typing. A highly discriminatory method results in more groups than a less discriminatory method. Discriminatory power is the average probability that a typing system will assign the same strain type to strains randomly sampled from the same group. Increasing the level of discrimination reduces the potential for misclassification, as it increases the homogeneity of a given group. However, the most discriminatory technique would put every individual into a separate group, which is useless. Thus it is equally important that there be some biological basis for collapsing categories to form larger groups. For typing techniques based on the genome, categories are generally collapsed on the basis of assumed genetic relatedness using phylogenetic techniques (see Chapter 7). The level of discriminatory power required depends on the study purpose (Table 9.7).

When there is epidemiologic evidence suggesting a relationship, such as when testing that multiple isolates from the same individual are the same or that epidemiologically linked cases from a disease outbreak carry the same strain, the required level of discrimination is relatively low. There is a high prior probability that the isolates will be the same; the molecular data are intended to either support or refute that contention. Because of the belief that outbreaks are usually caused by the same strain, validation of a typing technique is often done in the context of an outbreak. However, in this case – similar to statistical testing – accepting the hypothesis is more problematic than rejecting it. It is more difficult to definitively prove two strains are the same than different. Microbes are biological organisms; as they reproduce, changes happen. The rate of change varies; RNA viruses in particular tend

**TABLE 9.7** Required Discriminatory Power and Need to Infer Genetic Relationships and/or Population Structure for Various Epidemiologic Applications of Molecular Typing Techniques

| Purpose | Example Research Goal | Discriminatory Power Needed | Need to Infer Genetic Relationships and/or Population Structure |
|---|---|---|---|
| Confirm epidemiologic linkage | • Determine if epidemiologically related cases share the identical organism. Result: either support or refute epidemiologic data. | Low | Low |
| Generate hypotheses about epidemiologic relationships between microbial strains in the absence of epidemiologic data | • Determine if time–space clustering surveillance isolates have identical or related genetic types. Result: trigger further epidemiologic investigation of related isolates.<br>• Determine if outbreak is propagated. Result: trigger investigation into how it is spread and/or control actions to stop spread.<br>• Relate clinical outcomes to strain types or to the presence of transferable genetic material, e.g., antimicrobial resistance on a plasmid. Result: improve patient care. | Moderate to high | Moderate |
| Describe distribution of microbial types and identify the determinants of that distribution | • Test the hypothesis of clonal spread versus independent origin of a particular strain over disparate geographic areas. Result: better predict emergence and spread of disease.<br>• Determine flow of infection from one group to another. Result: public health intervention.<br>• Identification of pathogenic factors. Result: develop new interventions or therapies specific to those factors. | Moderate to high | High |

*Source: Adapted, with permission, from Foxman et al.[3] (2005).*

to have high mutation rates. When stressed, bacteria are more likely to acquire genetic material via transformation. One stressor is exposure to antibiotics. Thus if a highly discriminatory method is used, the hypothesis might be erroneously rejected. However, if the method does not discriminate enough, the hypothesis might be erroneously accepted. For organisms where the biology is understood, it is possible to estimate the probability that mutations occur or that genetic material is acquired or lost, enabling the investigator to better interpret typing results. If there is not information on the range of types that might be expected outside of an outbreak, the investigator should be very cautious about any inferences regarding interpreting similarities or differences among strains, especially if there is no epidemiologic information available.

The population of a given microbe is very large and the number typed very small; this makes our ability to determine the prior probability that strains will be the same much poorer than in the absence of epidemiologic information. Therefore the selected typing method needs to be more discriminatory than one used in an outbreak situation. Space–time clusters can appear by chance alone. The probability that they occur depends on the frequency of the strain type within the population. As a more discriminatory technique identifies more groups, it will detect fewer clusters occurring by chance than a less discriminatory technique. One question posed during outbreaks is if the outbreak is confined to a geographic area or more widely disseminated. If there is a clear chain of transmission or suspected vehicle of transmission, a less discriminatory method is needed than if hypothesizing broad dissemination. The reasoning is the same: the need to minimize the probability that a particular strain might appear by chance.

Inferences from studies of microbial strains not collected from defined populations are just as problematic as inferences from any source of convenience sample. Our understanding of the distribution of types for any microbe is poor because the number typed in databases is very, very small compared to the number of microbes in the world. Even when sampling a pathogen from a single infected human, there is a population of microbes that, if all were sequenced, would undoubtedly show a degree of variation. Thus, studies intended to describe the distribution of microbial types and the determinants of that distribution are least subject to bias if there is a defined sampling frame. Since the epidemiologic linkages between isolates are unknown (with the exception of information from the sampling frame), a more discriminatory typing technique should be used. For example, one scientific question of clinical relevance is the extent that a more severe clinical presentation can be attributed to a particularly virulent strain of the infecting microbe. These studies usually collect strains from those with severe disease at a number of locations, say all hospitals within a geographic area. The typing techniques should be sufficiently discriminatory that apparent linkage is less likely attributable to chance. Moreover, it is essential that control strains be collected for comparison.

## References

1. D.J. Dlugos, T.M. Scattergood, T.N. Ferraro, et al., Recruitment rates and fear of phlebotomy in pediatric patients in a genetic study of epilepsy, Epilepsy Behav. 6 (3) (2005) 444–446.

2. N.T. Holland, L. Pfleger, E. Berger, et al., Molecular epidemiology biomarkers – sample collection and processing considerations, Toxicol. Appl. Pharmacol. 206 (2) (2005) 261–268.

3. B. Foxman, L. Zhang, J. Koopman, et al., Choosing an appropriate bacterial typing technique for epidemiologic studies, Epidemiol. Perspect. Innov. 2 (1) (2005) 10.

# CHAPTER 10

# Study Conduct

Study results are only as good as the data that are collected, thus it is essential to be as meticulous as possible in collecting, storing, and recording data. The goal of study conduct is to maintain consistency in the myriad of details surrounding the recruitment, retention, collection, and processing of data from study participants so that each data point varies only because of true differences in the participant characteristics of interest – not because of differences in study conduct. It is inevitable that over the course of data collection that there will be changes in study conduct. As personnel become better acquainted with the study population and collection site they will modify their procedures to increase response rate and study efficiency. Study personnel or personnel in the field site that refer study participants may turn over. Collection devices, laboratory supplies, or test kits may be modified by the manufacturers. Software and hardware may be updated. Each of these changes may seem negligible, but the cumulative effect may be significant. Thus, while it is probably impossible to exactly maintain consistency across the study, there are many strategies that aid the investigator to come close, by thoroughly documenting all protocols and operating procedures, training and retraining personnel in all protocols and procedures, detecting and correcting errors, and monitoring changes over time (Table 10.1).

## 10.1 DOCUMENTATION OF ALL PROTOCOLS AND OPERATING PROCEDURES

Unlike clinical trials where there are a number of regulations regarding appropriate documentation and many templates with associated software available, requirements for documentation in observational studies are minimal or nonexistent. Nonetheless, documentation is essential, and only the most naïve investigator would believe that they can accurately remember all the decisions made in the process of planning and fielding an epidemiologic study. Thus the investigator should record the reasoning that led to a particular protocol or procedure, as well as when and why changes were implemented. All protocols and procedures should be documented, and the documents made available to all personnel as a reference. Documentation provides the basis for training study personnel and evaluating performance of personnel and study sites. It also serves as a reference for writing up reports and manuscripts for publication. Documentation also clarifies personnel and unit roles; this enables the principal investigator to make sure that all tasks are assigned and know which individual or unit is responsible if there is a problem. Further, documentation can codify usual procedures for such contentious issues as authorship on papers; by clarifying these up front, many problems can be avoided.

Documentation for a study is generally consolidated in a single study manual. The study manual should include all forms used in data collection, codebooks, descriptions of data flow, certifications, and quality control and quality assurance procedures (Table 10.2). Studies tend to develop their own shorthand for referring to procedures or specific variables. These should all be defined in the study dictionary or codebooks to avoid confusion as personnel change or after the study is completed. The study manual should also include a section to

145

Molecular Tools and Infectious Disease Epidemiology.
© 2012 Elsevier Inc. All rights reserved.

**TABLE 10.1** Strategies to Minimize and Monitor Errors in Study Conduct

- Documentation of all protocols and operating procedures
- Regular meetings with study personnel
- Training and retraining of study personnel
- Quality control and quality assurance procedures
- Interim data analyses

**TABLE 10.2** Documents Typically Included in the Study Manual for Molecular Epidemiologic Studies

- Study objectives, rationale, and design
- Study protocol (inclusion/exclusion criteria, study schedule, data flow)
- Informed consent form
- Data collection forms (questionnaires, record review forms, examinations, etc.)
- Data tracking forms (checklists for data collection, bills of lading, etc.)
- Codebooks and dictionaries of study jargon
- Laboratory protocols (specimen flow, laboratory procedures, etc.)
- Certifications (laboratory certifications, pledges of confidentiality, licenses, etc.)
- Quality control/quality assurance plan
- Notes regarding changes

note any deviations to the protocol and how they were handled. If the protocol is changed, the study manual should note the date of the change, and the new protocol should be added. The old protocol should remain in the study manual, but be clearly labeled as discontinued, noting the date it was discontinued. New protocols should be implemented during a period when the existing protocol is still in use, so that if there are any systematic differences in the resulting data, results can be adjusted during analysis. The date is important so differences in data before and after the change can be assessed. Study manuals are essential for training and retraining of personnel, ensuring consistency in handling protocol deviations, and for writing up study methods and analyzing study results.

All personnel should have easy access to the study manual. Because study manuals are "living" documents, keeping the manual in web-based or electronic format is desirable, especially if personnel have easy access to computers. The date the document was last updated, the version number, or both should be clearly indicated. Only select persons should be able to update the study manual, and only after following some clearly delineated process. Though the goal is to make study conduct similar as possible among study personnel and study sites, it is essential that the actual and not the hypothetical procedures are reflected in the study manual. If there are site differences, comparisons of the study procedures can give some insight into possible explanations. Electronic documents should be backed up regularly. Print copies may be useful for laboratory settings, as they may be easier to accommodate at the lab bench. Print copies also should be updated regularly, with the date of printing on each page. If print copies are used, care should be taken that all copies are updated when the protocol is amended.

Although study manuals should be updated frequently, they also should function as a diary or history of the study procedures. Thus if a new protocol is implemented, the older one should be kept within the manual but marked as no longer in use. The date it was discontinued and the name, location, and date of the replacement document should be included. Study personnel tend to make notes in their study manuals, clarifying procedures or noting modifications. Personnel should be made to feel comfortable sharing these notes with the principal investigator, and procedures should be developed for determining when these notes constitute a true modification that should be noted in the study manual. If using a web-based system, it is useful to include a discussion section for a running dialogue on study procedures; personnel should be encouraged to use this forum rather than email for discussing issues related to the study. Minutes of regular meetings might also be posted. These documents can be an important reference for the rationale behind various decisions that may be difficult to recall after the study is completed.

Study documentation should also include a description of the types of training and certification required for each job position. There also may be records of documents that must be signed, such as statements of confidentiality, or statements certifying that a staff member was offered (and accepted or refused) vaccination or similar.

Even a small study should be well documented. Documentation assists in the data analysis by providing dates of protocol changes for analytical exploration, data interpretation, and write-up for publication. If there is any question about how the research was conducted, or how a specific protocol deviation was handled, a meticulously kept study manual not only provides the answers, it also protects the investigator from a malicious lawsuit.

## 10.2 REGULAR MEETINGS WITH STUDY PERSONNEL

Regardless of the existence of detailed documentation, study personnel may or may not follow study protocols. The reasons may range from an unworkable protocol to poorly trained staff to a belief that the procedures don't make a difference. Most investigators have horror stories of discovering a research assistant had changed a procedure (shortened an incubation period to fit their schedule or didn't make follow-up phone calls) because he or she didn't believe it made a difference. Frequent meetings with study personnel, combined with quality control and quality assurance procedures help detect and correct these errors early. Ancillary personnel, such as medical personnel referring or recruiting participants, should also be contacted regularly. To be effective, the investigator should strive to create an atmosphere that makes personnel feel safe to raise concerns, and that their concerns are valued and will be addressed. The atmosphere should also underscore that ALL personnel contribute to the overall success of the scientific project. Just like in the story of the emperor's new clothes, it may be the most naïve individual who detects a fatal flaw or detects a novel pattern that leads to a totally new direction for study. If we knew the answers, a study wouldn't be worth conducting; thus the investigator should keep an open mind, and encourage all study personnel to do the same.

Regular meetings also smooth the transition of data from the field to the laboratory, and from the laboratory to the electronic database. Molecular epidemiologic studies generally include personnel with disparate expertise, perspectives, and goals. The goal of field staff is to recruit and retain study participants; laboratory personnel are concerned with integrity of specimens, specimen handling, and processing; and data managers are focused on data completeness. Though all concerned want as many samples as possible, laboratory and data personnel may be more sensitive to how poor data quality can make recruitment efforts for naught, because data from participants with poor data might have to be excluded. Regular meetings, as well as friendly social occasions, can smooth friction between these perspectives and help develop and maintain protocols that address the concerns of each.

## 10.3 TRAINING AND RETRAINING OF STUDY PERSONNEL

Training is an ongoing process (Table 10.3). Study personnel should be trained in study protocols before beginning any data collection (including pilot testing). After pilot study protocols and procedures have been tested, reliability and validity assessments completed, and documentation is in place, any study personnel who worked during the pilot phase and all new personnel should be trained in the final collection procedures. Whether there is one staff member or 1000, there should be a formal training period. For studies lasting for

147

**TABLE 10.3 Times to Train Personnel**

- Before beginning data collection (pilot phase)
- Before beginning data collection (final phase)
- Annually
- When new personnel are hired
- When the study protocol is modified
- If a problem is detected

several years, training should be conducted annually, to detect and correct the minor drifts in protocol that occur over time. In addition, all new personnel should undergo training, and if a problem is detected, all study personnel should be retrained.

Personnel invested in a study are more likely to adhere to study procedures. However, the investment must not be such that it influences study findings. Thus training should introduce staff to the problem under study, emphasizing the importance of strictly following the study protocol so the study results will be an objective evaluation of the hypothesis under study. Training should include a walk through the protocol, a practice period, and discussion. Staff might practice the protocol on each other; it is also useful during pilot testing to have field staff work in tandem, where one observes the other and gives feedback. The discussion period enables staff to raise concerns and collectively brainstorm about possible solutions. This is a good opportunity for staff from multisite studies to learn from each other and to come up with a protocol that is as similar as possible across study sites. It is also an opportunity to make staff aware of the roles of other parts of a group, and how the parts fit together. Field staff may be unaware of the extensive role of laboratory personnel and vice versa. Training might include presentations by laboratory personnel, who can highlight the importance of specimen collection and handling from a laboratory perspective; and field staff might present their specimen collection and handling protocols to laboratory personnel, highlighting the challenges. By sharing information and working together, effective strategies can be identified to meet concerns from a field and laboratory perspective.

Staff training should go beyond the specific study protocol. All staff should be trained in responsible conduct of research, protection of human subjects, and application of the Health Insurance Portability and Accountability Act and any other relevant rules and regulations. These regulations may include training in biosafety, proper shipping procedures for specimens, and in Occupational Safety and Health Administration regulations. Institutional, state, or federal regulations may require that study personnel undergo specific training. If study personnel will be handling specimens that may contain infectious agents (such as hepatitis) for which vaccines are available, it may be required that they be offered the vaccine. Ultimately the principal investigator is responsible if regulations are not followed.

In addition to training in the conduct of all study experiments, laboratory training should include introduction to the use and maintenance of equipment, use of lab notebooks, study databases and data entry procedures, and laboratory safety. If specimens are to be shipped to other institutions, training (and documentation) should include how to complete the appropriate paperwork, and packing and shipping procedures. Finally, laboratory personnel should be encouraged to record observations beyond those required for the study at hand. An astute technician may observe a new phenotype or unexpected organism or phenomena that can lead to the development of additional studies.

## 10.4 QUALITY CONTROL AND QUALITY ASSURANCE

Ensuring that data collection and processing is as consistent as possible over time and space requires instituting procedures to ensure that data collection and processing meets a specific standard of quality (quality control), and that there are ongoing procedures to

| TABLE 10.4 Quality Control and Quality Assurance Procedures |
|---|

- Verifying rights and welfare of participants and health and safety of study personnel are protected
- Ongoing monitoring of data quality
  - Specimen quantity
  - Data completeness
  - Accuracy and precision of measuring instruments
  - Detecting and preventing variations over time and space
- Repeat laboratory determinations/duplicate data entry
- Regular calibration of instruments
- Feedback for corrective action

monitor, verify, and document performance (quality assurance). Examples of quality control procedures in molecular epidemiologic studies include setting inclusion and exclusion criteria for participants, including positive and negative controls in laboratory tests, running tests in duplicate or triplicate, and including results only when the variation between replicates is less than some set point. To have results that meet these standards requires quality assurance procedures. These procedures range from verifying that participants' rights and welfare and the health and safety of study personnel are protected, to regularly calibrating instruments, to verifying that the recorded data is accurate via duplicate data coding and entry. While maintaining data quality is a concern in all studies, molecular epidemiologic studies have the added challenges of specimen collection, handling, testing, and storage; more complex data management; ensuring data quality from multiple sources; and detecting and preventing variations in data over time and space (Table 10.4). There is an entire literature devoted to quality control and quality assurance procedures, so only an overview will be presented here. For a more in-depth presentation of quality control and quality assurance, the interested reader is referred to the many textbooks on the subject.

Quality control and quality assurance requires ongoing monitoring of data. The purpose of monitoring is to maintain the completeness, accuracy, and precision of data, and to detect and prevent variations over time and space due to technical errors. Monitoring entails setting up systems and procedures to collect the data needed for monitoring, analyzing the data, and using the results of the analysis to correct and improve study conduct. The data needed for monitoring depends on the study but should enable the investigator to identify when and where an error took place. Thus the date, time, and location when specimens were collected, shipped to the laboratory, and received at the laboratory, the amount and quality of sample received, and so on are all important. The initials of the person conducting the procedure should also be included. In addition, equipment used for the study should be checked and calibrated regularly, and any concerns noted. For example, fluctuations in freezer and refrigerator temperatures may adversely affect sample quality, and improperly calibrated pipets will adversely affect test results. All electronic systems should be backed up by having a person check them at regular intervals.

Monitoring of data quality includes checking forms for completeness, ensuring that specimens were collected and in sufficient amounts, replicating experiments and data entry, performing quality checks of data for out-of-range values, and data queries regarding missing or incomplete data. Study protocols should include periodic check points in data collection, handling, and testing to minimize data losses, for example, forms to track that all forms are completed and specimens collected during each patient contact, and logs listing all items transported between the field site and data collection center or laboratory. Tracking forms should be analyzed regularly to determine how many and where data losses occur; if the protocol is modified to address the problem, analysis can determine if the modification was effective. Ongoing data monitoring is also essential to identify any secular trends in the data. Secular trends may indicate changes in how the study is being conducted; different characteristics of study participants; addition of new staff, equipment, or reagents; or a trend occurring outside the study. Secular trends can be minimized by testing specimens in an order independent of treatment assignment, exposure, and case or control status, and masking personnel, to the extent possible, to the same.

Secular trends may indicate assay drift. Assay drift is a phenomenon where a laboratory method gives increasingly higher or lower results over time. Assays that are done intermittently are particularly subject to drift; personnel have less experience and the technical variation may be greater. Drifts can also occur if the assay is sensitive to storage conditions; for example, the controls or standards may drift in value over time with storage. Use of standards provides a check on assay drift, but this check is overcome if the standards themselves drift with time. If samples are batched for testing, and there is a modest drift in value with storage, those collected earlier will test lower than those collected later. This can be avoided by testing samples as collected or batching over a shorter time period; however, this may be a more costly option. If specimens are grouped by an important variable under study, for example, case, control, or treatment status, assay drift can lead to an erroneous association. For example, imagine there is a downward drift in results over time, say only 2% per year. Cases are enrolled and tested in years 1 and 2, then controls in years 3 and 4. Even if there is no real difference between the groups, the cases will have higher values because of the drift in assay results. Because there is run-to-run variation for any assay, including only cases, only controls, or only samples with one level of important variable of interest for testing in a single assay should be avoided.

To monitor for assay drift, control values should be plotted over time; a standard method is using a Levey-Jennings chart (see Figure 10.1). A Levey-Jennings chart graphs control values per run or per day on a chart where expected mean value and lines showing plus or minus one, two, and three standard deviations of that value are noted. The variance between runs can be identified by the scatter of the control values around the mean; the closer to the

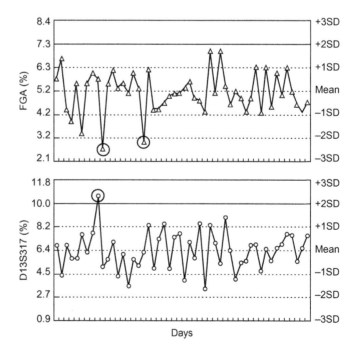

**FIGURE 10.1**

Example of Levey-Jennings chart from a study of microsatellites in allogeneic bone marrow transplant patients. Tick marks on the x-axis represent successive days on which the assay was run. Circles indicate control values that exceeded 2 SDs. *Source: Reproduced, with permission, from Liang et al.[1] (2008).*

mean, the more precise. A trend either upward or downward suggests assay drift. The graph is also used to decide whether a run is valid; if control values fall too far outside of the expected range, the run is suspect.

In the example shown in Figure 10.1, Liang and colleagues[1] (2008) used microsatellites to detect percentage of engraftment; to determine the quality control parameters, they used results from 20 runs of a sample from two individuals with one arbitrarily assigned as donor, with a ratio of 95% donor and 5% recipient, a clinically relevant cutpoint. Days are shown on the *x*-axis and the percentage of engraftment of the allogeneic donor samples is shown on the *y*-axis. There are two different controls, one for each microsatellite (FGA and D13S317), which are graphed separately. Ideally, all control values should be close to the mean, and there should be no trend in the values. There are various decision rules, but a standard one is that the control value should be within two standard deviations from the mean; values falling outside that range are cause for investigation. In Figure 10.1 there are three such events, shown in circles. Possible explanations are changes in testing procedures or equipment, equipment malfunction or in need of recalibration, change in personnel conducting the test, or drift in technique over time. The response to personnel changes or technique drift is to have a retraining session.

Developing a quality control and quality assurance plan requires outlining all steps from recruitment of study participants through data analysis and identifying the potential for errors. A target for data quality should be set (quality control) and procedures identified for monitoring the data to ensure that the desired quality is achieved. If the data quality does not meet the targeted standards, corrective measures should be implemented. For example, Table 8.2 in Chapter 8 lists steps and potential errors for specimen collection for a study of group B *Streptococcus*; Table 10.5 expands this list to include quality control and quality

**TABLE 10.5** Example of Listing Data Handling and Processing Steps: Specimen Collection for Study of Group B *Streptococcus* (GBS) and Quality Control and Quality Assurance Procedures

| Procedure | Potential for Error | Quality Control | Quality Assurance |
|---|---|---|---|
| Self-collection of rectal specimen using a swab, placed into transport media | Specimen does not contain fecal matter | Ensure that all positive cultures are detected, and all negative cultures are truly negative | • Visually check swab for presence of fecal matter<br>• Culture swab on blood agar to detect whether fecal matter is present |
| | Swab placed in wrong vial | Collected swabs transported according to protocol | • Color code swabs and vials so a visual check will reveal whether it is in correct vial |
| | Swab not labeled or improperly labeled | All swabs are labeled correctly | • Swabs and vials are prelabeled<br>• Visual check that labeling is in place before transport for processing |
| | Specimen not collected | Adequate specimens are collected from all study participants | • Clear instructions, with checklist, for person collecting specimen<br>• Check that all specimens are collected before transport for processing<br>• Process for noting participant refusal to complete protocol |
| | Transport media improperly prepared or outdated | Use transport media that is within freshness dates | • Regular checks on supplies at data collection site(s), including checks that media are not outdated |

*(Continued)*

**TABLE 10.5** (Continued)

| Procedure | Potential for Error | Quality Control | Quality Assurance |
|---|---|---|---|
| Transport of specimen to laboratory | Specimen lost during transport to laboratory | All specimens transported between collection sites and laboratory | • Logging of all specimens collected<br>• Bills of lading to track transport of specimens, logging of specimens on receipt at processing points |
| | Labeling lost during transport | All specimens labeled | • Use of labels resistant to loss during transport conditions |
| | Specimen heated or cooled during transport | Specimens stored at proper temperature during transport | • Specimens insulated during transport |
| | Delays or inconsistent transport time | Monitoring of transport time | • Training of personnel |
| Culture specimen for GBS | Error in labeling | All specimens labeled correctly | • Training of personnel |
| | Break in sterile technique | All specimens handled using sterile technique | • Training of personnel<br>• Monitoring for contamination |
| | Culture too long/short, incorrect temperature or conditions | Specimens cultured according to protocol | • Training of personnel<br>• Written protocols<br>• Regular calibration of incubators |
| Identification of GBS | Error in identification | GBS isolates will be identified according to protocol | • Training of personnel<br>• Written protocols<br>• Verification of sample of isolates |
| | GBS isolate grown on plate, but not selected for further testing (identification) | All presumptive GBS isolates will be selected for further testing | • Training of personnel<br>• Written protocols |
| | Incorrect recording of results | All results recorded accurately | • Training of personnel<br>• Verification of recorded results<br>• Written protocols |
| | Break in sterile technique | All specimens handled using sterile technique | • Training of personnel<br>• Monitoring for contamination |
| Storage of isolates | Error in labeling | All specimens labeled according to protocol | • Training of personnel |
| | Storage media not properly prepared | Media prepared according to protocol | • Training of personnel<br>• Written protocols |
| | Break in sterile technique | All specimens handled using sterile technique | • Training of personnel<br>• Monitoring for contamination |
| | Storage at incorrect temperature, or change in temperature that impacts isolate integrity | Specimens stored at correct temperature | • Training of personnel<br>• Regular monitoring of storage temperature<br>• Alarms indicating temperature changes greater than average variance<br>• Written protocols |
| | Storage location not recorded or recorded incorrectly | Storage location for all isolates recorded as per protocol | • Training of personnel<br>• Written protocols |

assurance procedures. Note that there is a quality assurance step to detect each potential error, and a targeted level for data quality (quality control). The first potential error is that the specimen does not contain fecal matter; this can be detected by visual inspection at time of collection. Sometimes, however, fecal matter may not be visible on the swab. We expect that all individuals will have *Escherichia coli* in their bowel flora, so culturing the swab for *E. coli* as well as in selective media required for group B *Streptococcus* enables the investigator to distinguish an improperly collected sample from a true negative.

## 10.5 SPECIMEN HANDLING AND STORAGE

Specimens are precious. Although human genes may be invariant, the epigenome, host immune system, and infectious agents are not, so we can never go back and get a specimen from an individual in exactly the same state he or she was at the time a previous specimen was collected. This is especially true in the case of colonization and infection with an infectious agent; both the host and agent change over time as they interact with each other. Thus specimen handling and storage procedures should be in place to minimize potential for errors due to loss or improper handling and maximize ability to process for intended uses (Table 10.6).

Preparing a packet that contains all the materials required during a participant visit helps ensure that the study protocol is followed, and all specimens are collected from all participants. A typical packet has a cover with the study identification number on the front that lists each step in the protocol, has a way to mark that the step was completed, and to note any deviations from protocol. The cover should be saved as part of the participant record either by entering the data into a database, by filing the original or a copy in the physical file, or both. Inside the packet might be the consent form, instructions for collecting specimens, preprinted specimen labels with the study identification number that can be applied to each specimen collection container as the protocol is explained to the study participant, or prelabeled specimen collection kits. Vouchers for parking and participant incentives might also be included. The packet also serves as a memory aide for the study protocol for both research staff and study participant.

A complex protocol is less likely to be followed correctly by either participants or staff. The investigator and research staff should walk through the study protocol to identify ways to simplify the protocol as much as possible. If the participant will move to different locations, these should be noted on the packet cover, numbered so the order is clear and perhaps noted on a map. Signs should be posted to help guide participants. If the participant must carry study materials, like the packet and collection containers, consider giving a tote bag with two pockets as an incentive for participation – one pocket might hold the enrollment packet, the other collection materials or the participants' belongings. If participants must return for follow-up visits, schedule the next appointment before they leave the current one, send email

---

**TABLE 10.6** Specimen Handling and Storage

- Use pre-prepared collection packets
- Prepare easy to follow instructions, with visuals, readily available to person collecting the specimens
- Ensure that each specimen is appropriately labeled and collected according to protocol
- Log each specimens as collected, and note any discrepancies from study protocol
- Log receipt to the laboratory and any variations in transport conditions
- Split samples
- Specimen labeling

---

and phone reminders, and minimize effort involved in returning, for example, by paying for parking or transport. Consider making home visits.

Specimens may be collected by the participant, study personnel, or medical staff. Regardless of who collects the specimen, there should be clear instructions. If specimens are self-collected, the study protocol should include teaching the participant how to collect the specimens using the instructions; drawings or pictures can minimize error. Study personnel and medical staff should be trained and given clear written and graphic instructions that are readily available whenever they collect specimens. If collecting multiple specimens, color code labels on the collection vials, and use the same colors on the instructions to minimize errors. Using preprinted labels with the study identification number helps minimize labeling errors.

To minimize specimen loss or damage, it is essential to put protocols in place to log and track every specimen. The investigator should be able to track whether all intended specimens were collected from a given study participant, and if there were any deviations from the study protocol. All research staff should be aware of the steps from specimen collection to storage; it is worth the effort to create a flow diagram for inclusion in study documentation and for posting by the lab bench or the study recruiter's desk. Tracking should include whether the specimen received the required set(s) of tests and whether the data were recorded. As noted in the previous section, analyzing tracking data is an important component of quality control and quality assurance procedures.

Study procedures should include recording of protocol deviations. Even the best staff make mistakes, and events happen that are out of staff control. For example, it is possible that specimens cannot be transported within the designated time limit because of traffic, leading to processing outside of the optimum time limit. This may or may not compromise the study results, but recording that there was a problem will help in interpretation later, and in any decisions to delete out-of-range data. Thus the investigator should encourage an atmosphere that focuses on fixing the problem and recording that it occurred rather than assigning blame. Monitoring the frequency of deviations is very useful. A frequently occurring problem might be avoided by a protocol or personnel change.

Increasingly we are moving to all electronic systems, which are not infallible. Electronic systems must be easy to access and intuitive to use, and have excellent technical support. Any barriers to use will result in loss or misrecording of data. Backups of the data are essential to minimize data loss, as is a backup system to record data if the system is not functioning. Further, it is crucial to use data entry systems and not spreadsheets to record data. Although spreadsheets are very easy to use, they lack the checks found in data entry systems, which minimize data loss. For example, it is easy to sort a spreadsheet by column so the linkage between row and column is destroyed; or to inadvertently change the number in one cell and not receive a prompt to verify the change.

Specimens are often split for use in different tests and for storage. Usually it is not the specimen but the results of testing or culturing the specimen that are of interest. Splitting or culturing a specimen results in additional specimens for storage and future testing. Some products are renewable resources, such as bacterial isolates, whereas others are fixed amounts, such as blood. Fixed amount specimens should be aliquoted, and replicates stored in different locations; renewable resources should be stored at least in duplicate. As freezers are not infallible, keep working and backup sets in different locations. All this duplication can result in large numbers of vials even from a relatively small study; a system to organize the freezer so the location of each specimen is known is extremely useful. Electronic tracking systems using barcoding are commercially available.

Storing specimens requires ongoing monitoring. Although freezers come with backup systems and alarms, they are no substitute for regular checking by study personnel. In our laboratory the backup freezer had a malfunction that did not result in an alarm, so the entire contents were lost. Because freezers do fail, the investigator is strongly encouraged to develop an emergency plan for short-term storage while a freezer is under repair, before it happens. Freezer contents may have to be distributed into multiple locations, underscoring the need to have all freezer contents well labeled and tracked so they can be relocated following freezer repair or replacement.

To ensure that specimen results can be accurately linked to the correct individual requires meticulous recordkeeping. Each specimen must be labeled in a way that has meaning in the laboratory that also allows linkage to other data sets. Barcoding is one option; storage tubes can be purchased that are barcoded so that it is unnecessary to apply an external label. (Labels can fall off during laboratory processing that includes water baths, freezing, and thawing). Barcode technology has fallen in price so is worth the investment for any large study. However, it is an investment, as there must be a sufficient number of barcode readers in the laboratory for the number of personnel and appropriate software for linking specimens and data. Further, personnel must be trained in the use of the software, barcoders, and label printing, and a system set up for frequent backups so data are not lost. No technology can replace meticulous recordkeeping, as laboratory processing often requires several steps and potentially moving into several different tubes. If not using barcodes, the investigator should develop a study identification number that enables linkage across different types of data and has meaning in the laboratory.

Study identification numbers have to be short enough to handwrite on laboratory vials, but be sufficiently informative to the laboratory technician. A format that has worked well for bacterial isolates is as follows: K100-1-01. The letter (K) designates the study collection; this enables laboratory personnel to quickly distinguish between specimens from different studies. The first three digits (100) designate the number of the study participant. This number appears on all data collected, enabling data linkage. The next number (1) designates the type of specimen from which the bacteria were isolated, for example, the urine, and the last two digits designate which of several isolates were collected from that site. So this would be participant 100 from study K, a urinary isolate, and the first isolate picked from the plate.

Despite best efforts, specimens will not be collected from all study participants, and some specimens will be lost or be of insufficient quantity or quality to be used for all desired tests. In addition, some tests will be inconclusive, for example, some bacterial strains may not be typeable using the selected system. The laboratory database should be constructed in a way that enables the investigator to determine the reason that data are missing; for example, there might be a separate variable that indicates status of the specimen. Meticulous recordkeeping and an appropriately constructed database that facilitates merging of data from various sources and clinical, laboratory, and medical records are essential.

## 10.6 INTERIM DATA ANALYSIS

Interim data analyses are an important adjunct to quality assurance procedures. Many interim data analyses are conducted on performance measures, such as protocol adherence, which are collected through tracking and monitoring (Table 10.7). Other analyses assess data quality and quantity. Additionally, there is a need to look at the data itself to detect early trends that might result in additional studies or modification of data collection procedures. For some studies, such as randomized controlled double-blinded clinical trials, there may be specific protocols to monitor the data for adverse events or a clear effect that may lead to stopping the trial early. Like all surveillance activities, it is not looking at the data that is important, but feeding back the results of the analyses to those who can effect appropriate changes.

It is essential that study personnel follow study protocols. Deviations from the protocol can seriously compromise study findings. Thus it is critical to monitor for adherence to

**TABLE 10.7** Interim Data Analyses

- Protocol adherence
- Metrics of laboratory performance
- Metrics of clinic performance
- Metrics of staff performance
- Data quantity and quality
- Data trends
- Evaluating the effects of protocol changes

the study protocol, and take appropriate steps to correct protocol deviations. The first step in an epidemiologic study is screening for eligibility; a protocol might specify that all individuals with a certain diagnosis be screened. A simple comparison of the number of screening forms to those with the diagnosis at the study site can detect a protocol deviation; for example, study personnel may have only approached those who seemed likely to respond. The data on the screening form should be analyzed to determine the reasons for refusal and the characteristics of those who enroll. Not only does it help in identifying ways to prevent refusals of eligibles, it can detect fraud on the part of study personnel. It is very difficult to make up data that has the same degree of variability observed in reality. Monitoring tracking forms can identify where losses occur or when the study protocol is not followed. For example, the analysis may identify study sites that fail to collect certain specimens or do not label them properly or whose specimens get lost in transit.

Interim analyses also give metrics of laboratory, clinic, and staff performance. Tallies of specimens processed, interviews conducted, refusal rates per interviewer, and number of participants enrolled all give insight into how well staff are performing and if certain staff or study sites need additional attention or training. Comparing proportions recruited over those screened by recruiter can identify the most successful recruiters; these individuals should be tapped as presenters for training and retraining sessions. Tallies might also be made of whether reports are prepared or prepared in a timely manner, and the number of abstracts submitted to national meetings or manuscripts submitted. These measures of data quantity also help the investigator in determining if the study will meet proposed targets and if additional study sites must be added to enroll participants or additional staff to process specimens.

Meeting enrollment targets is important, but quantity is useless if the data are not of sufficient quality. Interim data analyses also are used to monitor data quality. For example, the coefficient of variation for laboratory analyses and the percentage of missing data in laboratory analyses, questionnaires, forms, and follow-up studies are all measures of quality. To validate data quality, a percentage of specimens should be retested blindly and the data for individuals checked for internal consistency. For example, the coefficient of variation should be less than some predefined value (usually less than 5%). To identify laboratory technicians needing retraining, Levey-Jennings graphs (see Figure 10.1) might be plotted by a technician. When validating questionnaire data, analyses can determine if the data are internally consistent. For example, it is unlikely that an individual who is married would report never having engaged in sexual activity.

Preliminary analysis to determine data trends can be extremely useful. A construct measured by a single variable might be strongly correlated with outcome; an interim analysis would detect this trend and enable the investigator to add additional variables to evaluate the meaning of the association. Serendipitous findings, such as noting the frequent appearance of an unexpected infectious agent in the laboratory, may lead to new studies or changes in procedures to limit contamination.

**TABLE 10.8** Evaluating a Protocol Change

- Study designs
  - Pre–post comparison
  - Crossover design
  - Run new and old protocols in parallel
- Analyses
  - Comparison of performance measures
  - Agreement between new and old methods

Protocols change over time either by design or out of necessity. An interim analysis might detect that the protocol that worked so well during the pilot phase did not scale up satisfactorily and requires changing, or the manufacturer of the kit used for laboratory assessments might send a notice that there will be a new and improved version and the old one discontinued. In either case, the investigator should implement a formal process to change the protocols and evaluate the impact of the change. Depending on the situation, different evaluation strategies might be used (Table 10.8). In the case of a relatively minor protocol change, a pre–post comparison is appropriate. Consider the case where the protocol worked out during the pilot but did not seem to scale up satisfactorily. There are several reasons that this might occur; one might be that personnel added since the pilot are not as effective as those used in the pilot phase. An interim analysis can assess individual performance; should it be that new personnel are not as effective they can be retrained. The effects of retraining can be evaluated by comparing pretraining and posttraining statistics.

Although a pre–post comparison is easy to implement, it is a weak study design because it does not control for other changes that might have taken place during the same time period. A crossover design is more powerful. Imagine that a significant correction in protocol is necessary and there are two possible alternatives. For a study with multiple sites, each site can be randomly assigned to start with one protocol, then after a set period switched to the other protocol. This enables the investigator to assess the impact of order; say that sites using protocol A then B, do worse than those using B then A. It also compensates for any other changes that might have taken place during the same time period. While either of these designs might be used for a change in laboratory procedure, if the protocol change involves using new reagents or testing methods the usual design is to split samples and run the tests in parallel.

There are two general classes of analyses for these types of evaluations (Table 10.8). The intention of the comparison of performance measures is to determine if the change in protocol resulted in an improvement in performance. For example, the investigator might compare the number of participants screened and enrolled in the study during the preperiod and postperiod. Difference between time periods might be tested using simple statistics such as chi square for categorical variables or $t$-tests for continuous variables. (These statistics are readily calculated using most spreadsheets and statistical packages.) By contrast, the second class of analyses is to determine agreement, for example, that the new laboratory test kit gives results that are comparable to the old one. Here comparisons are made using measures of agreement such as the kappa statistic, or assessments of the validity of the new measure relative to the old one, such as sensitivity and specificity. The evaluation of a new test kit also includes mapping the values of the new test relative to the old; this is essential so that the test results can be adjusted for the change in any analyses of the entire set.

## 10.7 TYPES OF INTERIM DATA ANALYSIS

There are several different types of analyses conducted throughout a study: interim analyses for quality assurance, interim analyses at the request of the granter or to detect early trends

**TABLE 10.9** Data Analyses

- Checks of data integrity
  - Data points fall within established ranges
  - Skip patterns are followed
  - Data from different sources match
  - Coefficient of variation for laboratory measures
  - Intraobserver and interobserver variation
- Descriptive analyses
  - Parameter estimates
  - Variations over time and space
- Analytic analyses
  - Associations of putative exposures with outcomes
  - Check for confounding and interaction
  - Data reduction

that might result in additional studies or modifying data collection procedures, final reports, and analyses on specific topics for manuscripts (Table 10.9). Regardless of the reason for analysis, the data analysis process should be subject to the same standards of conduct as other aspects of study conduct. The analysis should be thoroughly documented, the analyses should be repeatable, and analyses of variables with high levels of variability or large numbers of missing values should be interpreted with great caution. When a manuscript or report is completed, the analysis program should be rerun so that each number can be checked against the computer output, and the final run, including log files and data files, archived. What follows is a brief overview of the different types of analyses to give some guidance into how to approach the data from an epidemiologic perspective.

The first step in data analysis is to check for data integrity; this is often called data cleaning. All incoming forms should be reviewed for completeness; if data are missing, and cannot be completed, the reason data are missing should be noted. Forms should be scanned for obvious errors, such as data put in the wrong spot or transposition errors. A procedure for checking for obvious errors should be in place. If the records are changed, the person changing the data should initial the correction and a note made of why a change was made. (If on paper, a line might be drawn through the original and the correction made next to it and initialed. If electronic, a textual note giving the original value could be added.) This might involve checking with the interviewer and determining that the data were in fact put in the wrong spot. It is best to be highly conservative in making changes to the data, because there is always the possibility that the error reflects a true value. The data analyzed should accurately reflect the original study records. Any discrepancies between the data file and study records can raise questions regarding the integrity of study conduct.

The data are what the data are; the investigator should avoid correcting the actual data. Any corrections should be done as statements in the analytical program rather than to the data in the database; the reasons for correction should be noted in the program. This ensures that the database reflects what was actually collected. For example, outlier values may be removed from analysis, and the analysis limited to the subset that meets all eligibility criteria. However, the excluded data should be retrievable for future analyses or to address any challenges to the study that may occur.

Data are increasingly directly entered into the computer during data collection; laboratory instruments may produce a data file. All laboratory data included should be from runs where control values fall within acceptable ranges. Automated systems can do much to limit data entry errors, for example, checking that values are within range, and enforcing skip patterns in questionnaires. But the quality is only as good as the people who write the program and enter the data. Ideally, data transferred from paper copy to the computer are 100% verified;

this means that the data are entered twice, the files compared, and any discrepancies are checked against the original form. There are several ways of doing this, ranging from simply comparing the two files to a program that flags all inconsistencies when the data are entered a second time. If 100% verification is the norm, data entry errors are reduced from 3% to less than 1%. Personnel entering data should be able to override set limits, because ranges are based on best guesses and some real data may be out of range. Programs that enforce skip patterns should be vetted thoroughly as the program could enforce an incorrect skip pattern.

Once the data are on the computer, examining a simple frequency distribution can alert the researcher to errors. For example, duplicate records may be found by looking at the frequency of identification numbers. A simple program can be written to check for logic errors. Suppose that a "no" response to a "skip" question means the entire next section should be skipped. The program checks that the next section was indeed skipped, and if not, changes the data to conform. However, it is important to review the data frequencies before enforcing skips; if the numbers responding to the skip section do not correspond to the number answering "yes" to the trigger question, further investigation is required. The directions on the form may have been vague, so most respondents answered all the questions; alternatively, respondents might have been directed to skip to the wrong section.

Molecular epidemiologic studies require merging several different types of data: laboratory test results with clinical and epidemiologic data. Although some data types might be analyzed independently, the investigator must ensure that all participants are accounted for in each set, and all inclusionary and exclusionary criteria are consistently applied. Therefore an identifying number should be included on each record so all the data sets can be merged. Further, data structure is important. A database may have multiple observations per individual, say one per lab test, or put all lab tests on a single record. Although data management programs enable movement back and forth between these data structures, it is important to determine the unit of analysis for a particular study. For example, there may be multiple bacterial isolates collected per person; one analysis might consider each genetically or phenotypically unique isolate as the unit of analysis, and another might consider the individual as the unit of analysis. For the first analysis the data generally would be organized as one record per isolate; for the second, as one record per person. When merging or restructuring data, it is essential to check that the merger or restructuring has the correct number of records and that the appropriate records were merged together.

The status of every specimen and subsequent tests on that specimen should be easily ascertained from the data. If data are missing, the investigator should be able to determine why the data are missing. The data variable for each test result should include values that indicate the data are missing and an additional variable the reason why there are no data. Some reasons might be that the specimen was not collected, the specimen was lost, the specimen was of insufficient quantity or quality, or the assay results were inconclusive. Thus there should be a record for every eligible individual, even if not all specimens were collected. Data cleaning for laboratory data should include checks of completeness similar to that described for questionnaire data. If there is no specimen all laboratory results for that individual should reflect the lack of specimen. Additionally, cleaning should ascertain that only data from experiments where control values fell within acceptable ranges are included.

The investigator should be able to show what happened to each individual contacted for the study, if they refused, were excluded, or were lost to follow-up; or if protocols were completed fully or only partially. This enables calculation of the response rate and follow-up rates. A flow diagram showing refusals, exclusions, drop outs, and so on, is extremely helpful in this regard. A variable should be created that accurately classifies participants by reason for exclusion, including whether all protocols were completed. The analysis should explore the extent that those refusing participation or not completing study protocols differ from participants, and if those differences should be taken into account during the interpretation of the results. A similar flow diagram can be created for specimens and associated tests, if only for the investigators' records.

It is rare that there is complete data on all subjects, so the sample size may vary from analysis to analysis. The investigator should be able to account for the reasons for this variation, and should explore if missing data varies systematically, potentially biasing study findings. If the number of missing data points is small (one rule of thumb is <5%), cases with missing data might be deleted, but the underlying assumption is that the missing data occur at random; this may or may not be a reasonable assumption. The reader is referred to the several excellent textbooks describing methods for imputing data.

An additional step for analyses of laboratory data may entail reconciling results of replicate assays. Having multiple replicates gives an estimate of the precision and also flags specimens requiring additional testing. For example, assuming the assay was done correctly, results of multiple values for a single specimen may agree with an acceptable level of variation, agree with an unacceptable level of variation, or disagree. If the assay is run with more than two replicates there are even more possibilities. The investigator should set up criteria regarding interpretation and when results should be repeated. Some laboratory results, for example, from a microarray, require extensive analysis to translate to values useful for estimating epidemiologic parameters, such as normalizing to control values in addition to checking values across replicates. Laboratory data thus require cleaning before merging with other data sets.

It can be hard to know when to stop data cleaning and begin analysis. The line between the two is very thin; often initial data analyses reveal need for more investigation and clarification of data values. Indeed, the first step of data analysis is congruent with data cleaning: data description. Data description entails examining the frequencies of all variables and simple descriptive statistics, such as means and standard errors, ranges, and histograms. These analyses help develop an intuition about the data and an understanding of its strengths and weaknesses.

Data description helps answer such questions as the following. Are there sufficient numbers in each category that the variable can be used as is, or should collapsing categories be considered? Does the group seem homogeneous or heterogeneous? Is the variable normally distributed or does it seem to fit some other distribution? Are there any outliers? Do the numbers add up? If there is a skip question, is it answered only by people who are supposed to? If there is no specimen, are all laboratory test results noted as missing? Do some questions/variables have a larger proportion of missing answers? Does this indicate a possible response bias? Do data vary over time and space? Although there is a tendency to rush through this initial phase of data analysis, the importance of getting a good feel for the data cannot be overestimated. A feel for the data acts as insurance against accepting results revealed by erroneous analyses, such as a strong association driven by a few outlier values in a multivariate model.

The second phase of data analysis is to examine associations between variables, cross-tabulating or correlating major outcome variables by all exposure variables, confounders, and modifiers. Both graphical and statistical analyses are useful. Although there is a tendency to look only at statistical significance, the effect size and direction of effect is most salient. These analyses help answer such questions as the following. Is there a regular dose-response relationship? Do you see evidence of a threshold? Would a scale transformation make the relationship more linear? Which exposure variables are most strongly associated with the outcome(s) of interest? With each other? Are the potential confounders and effect modifiers strongly associated with the outcome(s) of interest? Are the potential confounders and effect modifiers also associated with the exposure variables of interest?

After identifying the factors most strongly associated with the outcome(s) of interest, the next step is to check for potential confounding and effect modification, also known as interaction. Beginning with the exposure variable most strongly associated with the outcome of interest, consider each strong exposure variable as a potential confounder and/or modifier for all other exposure–outcome associations. Check for confounding and effect modification. These three-way analyses address such questions such as the following. Is an effect seen at all levels of the third variable? Are the joint effects of the third variable and exposure of interest additive, multiplicative, intermediate, or other? Is this third variable a confounder? Do two of the confounders seem to adjust for the same thing? Do the data fit my hypothesis(es)? Are my results supported by the literature? If I used a different analytical technique would I have a similar interpretation of the results?

The final step is to use statistical models or adjustment procedures. These analyses answer the following questions. Does adjustment make a difference? What implications do the observed joint effects have for targeting high-risk groups or further studies in the area? Are there other possible explanations for the observed results? The type of analyses presented in a publication depends on the story to be presented and expected audience, and, of course, what was found in the data. If, for example, adjustment procedures make no difference, presenting the crude associations is appropriate.

Regardless of analysis, it is critical that the analytical programs be internally documented. This means adding comments so that readers of the program can follow what analyses were conducted and how data were included or excluded. Although the analyst may have no trouble remembering details as to why several versions of a particular derived variable were developed, if he or she must return to a project even a few weeks later it may be difficult to remember. It also is useful to have code in the final program that will output the data used in developing all graphs and tables included in the publication by graph or table for future reference.

Increasingly, investigators are requested to prepare data files for public access; this requires that data are well documented and the individual aspects of a data set are as transparent as possible. For data to be uploaded on public databases there may be required formats and documentation; submission may be required for publication, for example, gene sequences. Good documentation for all aspects of study conduct helps ensure that a well-conducted study can continue to contribute to scientific understanding long after the initial investigation is completed.

## References

1. S.L. Liang, M.T. Lin, M.J. Hafez, et al., Application of traditional clinical pathology quality control techniques to molecular pathology, J. Mol. Diagn. 10 (2) (2008) 142–146.

# CHAPTER 11

# Think about Data Analysis When Planning a Study

Data analysis involves data cleaning, data description, statistical inference, and modeling. Data cleaning, the process of error detection and correction, is required for all types of data. Before cleaning, errors in sequence data can run as high as 2%. With cleaning, this drops to <1/100,000 base pairs. Data description involves exploring the distributions of variables, identifying how variables relate to each other, and if that relationship is modified by other variables. Data description can be done graphically or numerically. Statistical inference involves parameter estimation, functional estimation, and hypothesis testing. Modeling includes fitting statistical models to the data, developing mathematical models to describe the system, and using computers to simulate the system.

There is no single right way to analyze data, but there are wrong ones. A good data analysis is appropriate for the data, takes full advantage of the data strengths, and uses a variety of strategies to offset data weaknesses. The data analytical toolbox for molecular epidemiology is large and includes methods from a variety of different areas of scientific inquiry including epidemiology, biostatistics, phylogenetics, bioinformatics, and ecology. The depth and breadth of the molecular epidemiologic analytical toolbox is another indicator of the synergies that emerge when integrating molecular tools with epidemiologic methods, which is the overall theme of this book. Because molecular epidemiologic data can be analyzed using different strategies, and the data are collected at the level of the microbe, the individual, and population, the same data may result in several different types of publications. It is unlikely that an individual investigator will have sufficient expertise to conduct the entire range of potential analyses, but he or she should be able to read and make appropriate inferences from these analyses. Further, the investigator should have a vision of the scientific areas where the planned study will contribute and include collaborators with appropriate expertise on the research team.

Data analysis is too large and complex a topic to be covered in any detail here. Moreover, analyses of molecular epidemiologic data is an area under active development. Therefore it is not my intention to teach specific data analysis techniques. Instead, I present an overview of the range of analyses that might be applied in molecular epidemiologic studies. The chapter begins with a discussion of how to ensure during the design phase that the resulting data can be analyzed to address the study hypotheses. Planning the data analysis simultaneously with the study design ensures that the study design is appropriate and that all variables of interest are measured, including variables that assess factors that modify or confound the hypothesized relationships. Identifying modifying or confounding relationships also helps the investigator to estimate more accurately the sample size required to obtain statistical significance. The chapter continues with a brief discussion of data cleaning, data structure, and closes with some general analytic strategies for data at the molecular, individual, and population levels.

163

Molecular Tools and Infectious Disease Epidemiology.
© 2012 Elsevier Inc. All rights reserved.

## 11.1 DATA ANALYSIS AND STUDY DESIGN

How data can be analyzed is very much a function of study design and conduct. Some parameters cannot be estimated using certain study designs, for example, incidence of disease using a cross-sectional design (see Chapter 4). But if a variable has not been collected, it cannot be analyzed at all! Therefore it is prudent to plan the data analysis at the same time the study is designed. Whatever the intended analysis, it is useful to have a conceptual model of the relationships of interest, and the specific hypotheses or research questions (including any secondary hypotheses or questions) under study. If the study is designed to address the research question and conducted appropriately, the analysis is generally straightforward: there should be sufficient statistical power to address the study hypotheses or specific aims, and the analyses will fall directly from the hypotheses or aims.

While most investigators start with a research question or hypothesis, it is extremely useful to put that question or hypothesis in context of the system under study. The system might be how a pathogen causes disease, moves through populations, or causes an outbreak. Placing the research question in context of the overall system points out (potential) relationships among variables that should be measured to address more completely the research question of interest. Conceptual models are one way to do this. The model may be quite simple, such as:

If the system is that simple, then the investigator need only measure the exposure and the disease. In epidemiology (and biology) this is virtually never the case; there are always other variables that influence the relationship. This can be represented in a slightly more elaborate, but common conceptual model in epidemiology:

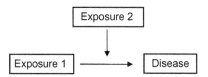

In this model, exposure 1 causes disease and exposure 2 modifies the effect of exposure 1 on disease risk. Eating food contaminated with *Salmonella* can lead to diarrhea, but the magnitude of the risk depends on the virulence characteristics of the particular strain of *Salmonella*, how much was consumed, and if the food was consumed alone or in the presence of some other food that modifies the bacterial growth. It also depends on the host. A person who takes antacids is at higher risk of disease, as is a person who has a condition that affects immune function. This conceptual model is often used during outbreak investigations: identify the source of infection and possible explanations for variation in disease rates among those exposed. Other common models used for outbreak investigations are the epidemiologic triangles of host–agent–environment, and person–place–time. Studies based on these models are sometimes referred to as risk factor or black box epidemiology, because there is little or no inclusion of the biological processes that occur as a result of the exposure that leads to disease. The focus is on inputs (exposure) and outputs (disease). They are also static; the outcomes are associations at single points in time rather than how disease changes with time.

At the population level there are well-developed epidemiologic models of disease spread, known as transmission models, which describe the dynamics of the population flow of infectious agents. Other dynamic models are based on evolutionary or ecological theory. Regardless of the theoretical basis, it is worthwhile to take the time to draw a simple conceptual model that demonstrates the relationship between theoretical relationships and the research questions at hand. Drawing the model requires making explicit the variables of

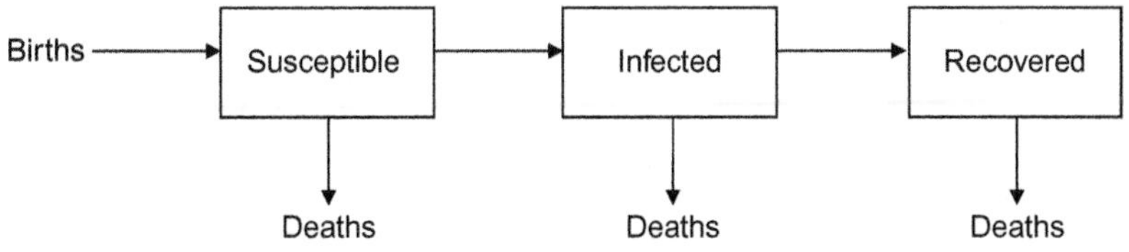

**FIGURE 11.1**
Example susceptible–infected–recovered (SIR) model. The rate that susceptible individuals become infected is the incidence rate; the rate that infected individuals recover is the recovery rate. Individuals are born susceptible and enter the population through birth, and leave the population through death, at a rate which varies by the compartment of the model.

interest and the relationships among them, and the scale(s) of interest. For a secondary data analysis, which uses already collected data, it helps clarify the strengths and limitations of using a particular data set to address the research question.

A common model used to describe an epidemic process is the susceptible–infected–removed (or recovered) (SIR) model (Figure 11.1). Although this and related models (such as susceptible–exposed–infectious–removed [SEIR] and susceptible-infected-susceptible [SIS]) are operationalized using mathematical equations to predict disease spread, they are also useful for conceptualizing a transmission system during the study design phase. Using this conceptual framework points the investigator to measure how infection flows through a population, using such parameters as transmission probabilities, duration of infectiousness, and length of immunity. This, in turn, points to specific molecular measures and study designs. Returning to the *Salmonella* example, using the SIR as a conceptual model might lead the investigator to describe the duration of asymptomatic human carriage and if it is sufficient to explain sporadic cases of *Salmonella* diarrhea not associated with specific foods.

Using an evolutionary framework focuses our attention on historical rather than proximate causes for what we observe: what past events best explain what we currently see? Evolution can occur because of natural selection, genetic drift, mutation, migration, and nonrandom mating. If the genetic code for a particular gene shows little variation across populations, the questions we might ask include, Was that gene under selection? What factors are selecting for its maintenance? We might also posit that a particular gene might have spread rapidly, for example, a horizontally transmitted gene for antibiotic resistance, so the reason for little variation might be that the gene itself is new. Ecological theory contributes yet another perspective. Rather than focusing on the effects of individual organisms or genes, it asks us to consider why we observe specific groups of organisms or genes in specific settings. What environmental factors select for specific microbial communities? Does the presence of one microbe interfere or enhance the ability of another to colonize? Whatever the conceptual model, the investigator should be able to show what part of the system under study is addressed by the study hypotheses. The combination of the conceptual model and study hypotheses provides a basis for the data analysis plan, which in turn gives a basis for developing lists of variables to be measured.

Once the data are collected, the data analysis plans should be considered as a road map, not as strict marching orders. We conduct studies to test hypotheses and describe phenomena; if we knew the answer(s), there would be no point in conducting the study. Further, epidemiologic studies are often conducted over long time periods during which our conceptualization and understanding of the problem may evolve. So, it is not unusual to find unexpected associations or for additional hypotheses to have been developed in the time from initiation to completion of data collection. Ancillary variables in the design phase may play major roles in the final write-up, and primary variables in the design phase may turn out to be uninformative. It is expected, and considered good scientific practice, even in a well-designed and well-conducted study, for the data analysis to include additional analyses that were not considered during the design phase. Moreover, it is a waste of an important resource if the data are not used to explore new or additional hypotheses beyond those originally considered.

Given the time and expense involved in conducting an epidemiologic study it is not unusual to conduct a secondary data analysis, in which the data are examined for a purpose not foreseen at the time the study was designed. Molecular tools increase opportunities for secondary analyses, because specimen repositories or even specimens that might have previously been discarded can be analyzed in new ways. For example, diagnostic blood specimens collected to measure cholesterol might be tested for antibodies to specific microbes. However, secondary analyses are constrained by all the design decisions made at the time of data collection. The investigator should consider whether the advantages of using already collected data outweigh the disadvantages.

## 11.2 ESTIMATING THE REQUIRED SAMPLE SIZE AND (CONVERSELY) DETERMINING IF THE CONDUCTED STUDY WAS SUFFICIENTLY LARGE

To determine the appropriate sample size for a particular analysis, the investigator makes a guess as to the size of the parameter of interest and desired level of statistical significance. If a study is planning to estimate the frequency of antibiotic resistance, the sample size required is quite different if the expected frequency is 1%, 5%, or 50%. Assuming the investigator wishes to estimate this proportion with precision of ±5% for a type I error of 5%, the sample sizes are 16, 73, and 384, respectively. But if the desired level of precision is 1%, which is much more sensible for estimating a value expected to be ~1%, the sample sizes are 381, 1822, and 9513. The required sample size is, however, only a guess; only after the study is conducted and the parameter is estimated is it clear that the sample size was sufficient. Once a study is completed or after the sample size is fixed, the investigator can estimate the power of the study to detect the parameter of interest or measure of effect at the desired level of type I error. The formula for estimating sample size is essentially the same as the ones used to test statistical significance; instead of solving for statistical significance the desired significance level (type I error) is included as part of the calculation and the formula is solved for the sample size. For power, the ability to detect a true association *if* it exists, the formula is solved for the beta value (type II error). Power is 1 minus the type II error.

Sample size formulas have four ingredients: the desired effect size, either a difference or ratio measure, the variance of the measure to be estimated, and the type I (alpha) and type II (beta) errors. A type I error occurs when one falsely rejects a true hypothesis; this is generally set at a level of .05 or below. A type I error corresponds to 1 − sensitivity or the false-positive rate, which is the probability of testing negative, given truly positive. A type II error occurs when one accepts a false hypothesis. A type II error corresponds to 1− specificity or the false-negative rate, which is the probability of testing positive, given truly negative. One minus the type II error is interpreted as the power of the study, or the probability of detecting an effect if it truly exists. The type II error is a function of the prevalence of exposure. Power increases as the ratio of exposed to unexposed (or for a case–control study, diseased to nondiseased) approaches 1:1. For sample size estimation the ratio of type I to type II errors is generally set as 1:4. There are different formulas for estimating sample size depending on what the investigator intends to test during the analysis; sample size calculators are freely available on the web (see for example, OpenEpi, http://www.openepi.com/).[1] (Regardless of the sophistication of the formula used, sample size estimation *is* an exercise in guesstimation.)

The required sample size decreases as the effect size increases: fewer numbers are needed to detect an odds ratio (OR) of 5 than one of 2. The greater the precision of the measure, the fewer numbers are needed, as there is less noise to be ruled out. The required sample size also decreases as the level of desired significance increases. A smaller sample is needed if the type I error is .05 rather than .01. Also, fewer numbers are needed if the groups to be compared are roughly equal in size. Consider a validation study that uses a case–control design. The investigator might choose an equal number of known positives and known negatives defined by some standard, and then compare the test results to "truth." A good test should give results very similar to truth, which means the effect size should be very large. If the test were perfect, the number of those testing positive would equal the number of cases, and the number testing negative would equal the number of controls. Placed in a $2 \times 2$ table and calculating an OR, the effect size would be infinite: the $c$ and $b$ cells would be zero, so the OR of $(a \times d)/(b \times c)$ has a zero in the denominator. Therefore the sample size can be relatively small and still show a statistically significant effect.

Suppose there is a new, rapid polymerase chain reaction (PCR) test for detecting the antibiotic resistance profile of bacteria causing bacterial pneumonia from nasopharyngeal swabs. The results are resistant or sensitive for the three antibiotics most commonly used to treat pneumonia. To validate the test, the investigator will compare results from culture of nasopharyngeal swabs from the same individual followed by antibiotic resistance testing to that of the PCR test. Cases will be selected from samples resistant to one or more of the antibiotics of interest and controls from samples universally sensitive. How many participants should be tested? If the test is perfect, not many. If we believe the test is a good one, we expect (almost) perfect sensitivity and specificity. What sample size is needed if 90% of the culture positives are positive by the test (sensitivity) and 90% of the culture negatives are negative by the test (specificity)? Depending on the sample size formula used, between 10 and 16 samples total!

Now consider validating a test where the result is a continuous rather than a dichotomous measure. Here, rather than measuring sensitivity and specificity, we estimate the correlation between the measures. For a dichotomous variable or proportion, like in the previous example, the variance is estimated based on the binomial distribution ($pq/n$, where $p$ is the proportion and $q$ is $1 - p$). Thus to calculate sample size the investigator must have an estimate of the proportion with the outcome among the unexposed (or proportion with exposure among those without the outcome for a case–control design). For a continuous variable, the investigator needs to know the distribution of the variable; for example, if it is approximately normally distributed, the investigator needs an estimate of the mean value and variance. For a correlation, variance is included in the calculation of the correlation. If we expect that the correlation coefficient will be high, say 90%, to reject the hypothesis that there is no correlation the sample size again will be small, ~5 subjects. Of course, for both validation studies, it would be best to include more subjects, and for the continuous measure, include samples over a range of values. Tests may function better in certain population subgroups or within a set range of values. Validation testing should be designed to capture the range of expected values and groups in the population of interest.

These sample sizes are quite small compared to what one might expect for nonvalidation studies. In the validation examples, the expected effect sizes are enormous: an OR of 81 for the dichotomous variable! For other studies, a more realistic effect size is an OR of 2.0. (Recall that odds ratios and risk ratios are on log scales, so the difference between 1 and 2 is greater than between 2 and 3.) In a case–control study, if 10% of controls are exposed, to detect an OR of 2.0 (with equal numbers of cases and controls), and alpha of .05 and beta of .20, we need between 284 and 307 individuals *per* group (depending on the formula used). If the outcome of interest is a continuous variable that is measured with precision, we can get away with a much smaller number. If we have a continuous variable where the mean value for controls is 10, with a standard deviation of 7, and the mean value for cases is 20, with the same standard deviation, the ratio of the two values is still 2. However, the sample size needed to test the difference in the two measures for an alpha of .05 and beta of .20 is only 8 per group. The sample size is the same as long as the expected difference and standard deviation are the same. Obviously, using a continuous measure and testing for differences greatly reduces the required sample size.

An additional consideration is what happens to an association in the presence of a modifying variable. That is, is there an interaction? Modifying variables requires more extensive analysis and a larger sample size. The larger sample size is needed because the analysis must consider what happens to the hypothesized effect in the presence and absence of the modifying variable (or across the range of the modifying variable). Thus the investigator should estimate the minimal sample size needed to detect an effect in these subanalyses. These estimations can lead to rethinking of the study inclusion criteria, because there is a cost in terms of recruitment, testing, and follow-up associated with each study participant. If the expected sample size for a particular subgroup will be insufficient to draw meaningful inferences, the investigator should consider other options. There are several options available: (1) the study can be restricted to groupings that are expected to be of meaningful size, (2) a study population in which where there is a more even distribution across the modifying variable can be selected, or (3) groups of particular interest can be oversampled.

Determining sample size for a validation study of a single measure or estimating sample size for a single hypothesis is quite straightforward. It is much harder to estimate the sample size for a molecular epidemiologic study where there are multiple outcomes and exposures of interest, and the analysis must consider the potentially modifying or confounding effects of third variables. The first difficulty is finding data to give an idea of the range of exposures and outcomes that might be expected. The expected range of values is required to calculate the desired sample size. Next, the investigator must guess at how large an effect might be expected. As this is often the reason for conducting the study in the first place, it is typical to identify a range of values that might be plausible and the range of sample sizes required. This process is repeated for all primary outcomes and exposures of interest. After the sample size is fixed for the primary effects of interest, the investigator might estimate the power of the study to detect secondary effects.

Estimating the study power *after* a study is conducted is also of interest if an unexpected finding emerges that is statistically significant. It is important to remember that one possible explanation for an observed association, even if statistically significant, is chance. One piece of evidence that can help evaluate whether an observed effect is real, in addition to examining the biological significance and checking for internal consistency of the effect in the data, is to calculate the power of the study to detect the effect. Imagine that in our case–control study, we enrolled 200 cases and 200 controls. Surprisingly, there is no association between exposure A and outcome, but 40% of cases and 30% of controls have exposure to A *and* B, giving an OR of 1.6. What is our power to detect this effect? For an alpha of 5%, the power to detect an OR of 1.6 is ~55% or a beta of 45%, meaning there is a 45% chance of falsely accepting a false hypothesis.

Ultimately every study runs into a small-sample size problem, because it is simply impractical and too costly to conduct a study large enough to estimate all effects of potential interest at the desired level of statistical significance. Given financial constraints, the investigator must choose a sample size based on the expected costs of recruitment, retention, specimen collection, handling, testing, and storage and restrict the study to the sample where the most meaningful results might be obtained. When justifying choice of sample size, it is essential to include the assumptions behind the estimate, which are the choice of alpha and beta, effect size, and variance of the measure.

## 11.3 DATA STRUCTURE

How the data are structured in the computer file makes a difference in the ease of data analysis. Although data can be manipulated, it can take considerable programming to transform data from one structure to another. Therefore I encourage you to take time up front to determine the optimal structure for the planned analyses. Imagine conducting a study where there are multiple tests per participant at each visit, and each participant

**TABLE 11.1** Examples of Different Types of Data Structure: All Results on One Record (Line) or Separate Records (Lines) by Visit Number

Identification_number date_visit1 test_results_visit1 date_visit2 test_results_visit2
Identification_number visit_number date_visit test_results
Identification_number visit_number date_visit test_results

has multiple visits. The data can be structured in several ways. One way is to place all information regarding an individual in a single record (line). Another way is to use multiple records per individual, where each record corresponds to a different visit. In Table 11.1 the first row shows the data all on one record (line); the variable containing information regarding the visit date must be labeled differently for each visit. In essence, the visit variable notes both the visit number and the date. This is also true for the test results. There must be a separate variable for each test by visit, for example, test at visit 1 and test at visit 2. By contrast the data in the second row have separated the two. There is a variable indicating the visit number and a variable for the date of visit, but each record (line) is a separate visit.

Data structure is important because many analytic programs require certain structures. When the data are all on one record (line) it is easy to use most statistical packages to perform analyses such as determining the number of days between the two visits, and using results of the test for visit 1 to predict those at visit 2 or the difference in test results between the visits. It will require some programming, however, to average the test results of the two visits within an individual or do any type of analysis where each visit is considered independent and equivalent. It is exactly these types of analyses where the second type of data structure excels. There are other reasons to choose one type of data structure over another. One reason is data entry. If the laboratory tests are conducted in a different location, or there are multiple tests over time, having a separate record per test eliminates the need to search for the appropriate record each time additional data are added. It also eliminates the potential for inadvertently entering data in the wrong place, such as visit 1 data in the visit 2 spot or vice versa, or mixing data across individuals. In practice, with most statistical packages it is relatively easy to manipulate the data from one structure into another as required. Therefore it is probably best to chose an initial structure that is easiest for data entry, keeping in mind that the given data structure might have to be manipulated to perform certain analyses. This means it is essential to have standardized identifying information on every record that can be used to merge data.

Molecular epidemiologic studies operate on multiple scales: that of the microbe, the host in which the microbe lives, and the population of hosts. At each level there are interactions among individuals in the population, and there are interactions between scales. The differences in scales operationalize into different data sets with different data structures. At the level of the microbe there might be one record per microbe with variables indicating the results of tests on that particular microbial strain. However, there might be multiple microbes isolated per host. For many analyses at the microbial level, it is easiest to have a separate record per microbe isolated per host, which includes a variable indicating the host identification number so that variables relating to the host can be merged with microbial characteristics. This data structure enables examining questions regarding microbial populations while analytically taking into account nonindependence of microbes isolated from the same host. However, if the purpose of the analysis is to determine how groups of microbes move together, for example, a cluster analysis, data on all microbes should be included on a single record (line).

If the unit of analysis is the human host, it is easiest to have one record per human. The relevant microbial variables, perhaps reduced to a summary measure, can be merged onto the record. For example, if the outcome of interest is whether *Escherichia coli* with a particular antibiotic resistance profile was isolated from a stool sample, a dichotomous (yes/no) variable indicating that was the case might appear on the record. That there were four other isolates tested without this profile would not appear. Alternatively, the data for each microbe tested might be included on the record. The second data structure enables examining questions regarding human populations and their interactions with microbial populations, including describing transmission of particular microbial strains among human hosts.

## 11.4  DATA CLEANING

Before beginning any data analysis the investigator should ensure that the data are clean. These types of data analyses are described in Chapter 10 as checks of data integrity. Briefly, data should be reviewed for completeness and accuracy before any inferences are made. In laboratory analyses, the positive and negative controls and any standards should fall within expected ranges. Sequence and other omics data should meet set criteria for quality. Criteria for including and excluding study participants should be checked for each record, the results recorded, and a determination made regarding the assignment of each record to the appropriate group(s) for analysis. In an epidemiologic study, data analysis is typically limited to all those meeting eligibility criteria or the inclusionary and exclusionary criteria. However, the eligibility criteria might vary for different analyses; for example, the enrollment group for a longitudinal sample may not all be eligible for follow-up, but may be an informative group for analysis on its own. Finally, if the data analyst is not the investigator, the analyst should spend the time required to become familiar with all aspects of the study, including the study protocol, the structure of all data files, the totality of data files available, the variables measured, and different types of measuring tools used, including their appropriate interpretation, strengths, and weaknesses. If working with a data analyst, the investigator should plan on meeting with the analyst regularly, reviewing interim reports, and discussing the analytical strategy to ensure that the data are appropriately interpreted.

In addition to setting up quality control and quality assessment procedures as described in Chapter 10, data cleaning can be facilitated by judicious choice of data values recorded. Because computers are sensitive to the case of text, using text values can considerably lengthen data cleaning. If studying antibiotic resistance it is much easier to record the number 0 as sensitive and 1 as resistant than it is to sort through text values of "Sens," "sens," "Se," "se," "Sen," "SEN," and so on, which all have the same meaning.

Even in the best of studies, some variables will have missing data points. There are many reasons why data might be missing: the study participant refused to answer a question or give a specimen; a specimen was lost or stored improperly so it cannot be used; for whatever reason, it was impossible to conduct the necessary test. Missing data are a problem. They reduce the effective sample size, decreasing the power of the study to detect an association. For multivariate analyses, any record with a missing data point in any of the variables included in the multivariate model is removed. If the data are not missing at random, the estimated parameters will be biased. Imagine that individuals with the most severe illness are least likely to be followed over time because they are too ill to participate. It will be difficult to accurately predict prognosis if these individuals are excluded. For these reasons it is essential to minimize missing data points, obtain ancillary information if available, and to document the reason that the variables are not measured. One rule of thumb is that if more than 5% of the data are missing, the measure is suspect, and if greater than 20% are missing, the measure is uninformative. There are, however, analytical strategies to impute missing values. One might assume that those with missing values have the mean value for the population, or impute a value based on the distribution of other values that are known for that individual. It is important to ensure that imputed values are conservative – any introduced biases should be toward the null hypothesis of no effect – and the analysis should include a sensitivity assessment. The sensitivity assessment might vary the imputed values to determine the degree that imputation influences the findings. For a thorough introduction to handling missing data, the reader is referred to textbooks and the statistical literature on missing data (see for example, *Statistical Analysis with Missing Data*, by Little and Rubin.[2])

## 11.5 GENERAL ANALYTIC STRATEGIES

Once the data are clean, they should be described. The investigator should take time to thoroughly explore the data using simple descriptive statistics, such as means, medians, ranges, variance, and standard deviations, and graphical and categorical analyses (Table 11.2). This exploration will quickly identify variables that are suspect because the variance is much greater than expected, many of the values are missing, or the results are inconsistent with related measures in a way that cannot be explained. It will identify variables that are highly correlated, giving the same information in a way that can cause multicollinearity problems if both are included in a regression model. It will also identify points outside the normal range (outliers), which should be double-checked and possibly excluded from analyses that are strongly influenced by outliers, such as linear regression. Exploring how similar measures are related assists in building derived variables, a single variable that reflects the best of multiple measures. The data should be internally consistent; if variables measuring different aspects of the same construct are not correlated the investigator should look further. Inconsistencies in what are anticipated to be correlated variables can identify errors in data collection or specimen testing or in the understanding of the underlying biological process. Most importantly, data description builds an intuition for the inherent strengths and weaknesses of the data, and of the relationships among the variables. As the analysis becomes more complex, using multivariate models and clustering procedures, the number of assumptions about the relationships among variables and the underlying distributions of the variables increases. Analytical techniques vary in how sensitive they are to the underlying assumptions. If the results conflict with the picture developed during the descriptive phase of analysis, they probably are wrong and the analyst should seek the source of the problem. This applies regardless of the research question or scale.

Some variables are measured using a scale that suggests a greater level of precision than is the reality given the variance and the relationship to biological function. For example, dot blot intensities are scanned, the background intensity subtracted and a value given based on the density of pixels observed. However, the purpose of the assay is generally to discern if a gene is present or absent; the density must be interpreted to make this determination. For a variety of technical reasons, such as the amount of DNA present, two spots with wildly different densities may give the same result. Therefore the interpretation of any test values should be established before any analysis using the appropriate positive and negative controls.

**TABLE 11.2** Data Description

- Examine the distributions of each variable
- Explore how different measures of the same characteristic are related
- Create derived variables, and examine their distributions
- Use categorical or graphical analysis to examine relationships among the known and hypothesized predictors of outcome
- Use categorical or graphical analysis to examine relationships among the various measures of the outcome of interest
- Use categorical or graphical analysis to examine relationship between known and hypothesized predictors and the outcome measure(s)
- Use categorical or graphical analysis to examine relationships among the known and hypothesized predictors and the outcome measure(s) stratified by potential modifiers of that relationship
- Estimate crude measures of association between known and hypothesized predictors and the outcome measure(s)

Finally, for analyses intended to make inferences regarding causal relationships, the analyst should have a good understanding of what constitutes the etiologically relevant period, which is the time when a causal factor is acting. One expects a factor that makes an individual more susceptible to infection to be present before the time of infection. Similarly, a factor modifying risk of disease should be present after colonization but before disease. Some factors do not vary with time, for example human genotype, but host immune response or host behaviors that influence disease risk, such as diet or sexual behavior, are often time dependent, and might change in response to infection. As data are analyzed and interpreted, the investigators should keep in mind when variables were measured relative to what is known about the biological process under study. Analyses focused on historical causes should accommodate the generation time of the microorganism and the molecular clock. The remaining sections discuss the types of analytical questions and some analytical strategies that might be used to address them from the molecular to population level.

## 11.6 MOLECULAR LEVEL

Epidemiologists are concerned with populations, most often human populations, but molecular epidemiologists also analyze microbial populations within a host, in the environment, and within human and other populations. Some of the associated research questions thus overlap with those of disease ecologists. Examples of the types of questions that might be asked at the molecular level are shown in Table 11.3. These questions focus on how microbes or specific microbial characteristics move through microbial populations or how and when that characteristic emerged.

Analyses of distributions by time and geographic area may use standard descriptive statistics; but the analysis often uses the techniques of phylogenetics to group microbes or genes that mutate over time. Phylogenetic techniques also can be used to estimate the genetic origin and to refine guesses as to order of mutation. Within the context of an outbreak investigation, phylogenetic analyses can give insight into the order of transmission. In some bacterial species, analyses of how strains vary over time and space have an additional challenge. Unlike animals, recombination and acquisition of new genetic material can occur in bacteria independent of reproduction. Gene transfer allows one microbe to have genes of different genetic origin; such microbes are known as recombinants.

Horizontal gene transfer in bacteria includes acquisition of genetic material from the same or different species via conjugation, infection by phage or virus that integrates into the genome (transduction), and direct uptake of DNA from the environment that is then integrated into the genome of the cell (transformation). In eukaryotes, recombination, that is, the joining of two different pieces of DNA, occurs during meiosis. Chromosomal crossover is a type of recombination. The generation time of microbes is considerably faster than that of humans, and some microbes, for example, RNA viruses, have high mutation rates. For studies of recombinants, the first step is determining which genes are of interest. Variations over space and time can be traced using markers of mutations, such as single nucleotide polymorphisms (SNPs), in specific genes or transposable elements.

There are two types of analyses for addressing questions regarding variations over space and time. One is similar to standard epidemiologic analyses – the proportion of genetic variation in one or multiple genetic loci is compared between different microbial

---

**TABLE 11.3 Analytic Questions at the Molecular Level about Individual Microbes**

- How do microbial strains or specific microbial genes vary over time and geographic area?
- What is the origin of a particular microbe or microbial characteristic?
- When did a microbial strain or species emerge?

populations. The microbial populations may be defined as those living within a single host, or groups of hosts related by genes or social or behavioral factors, geographic area, or a combination of variables. Differences in the proportion of the gene(s) or alleles of interest by population can be used to estimate the relative rates of transmission among the different populations and give insight into the potential pathways of transmission.[3] Related analyses might examine whether a gene is highly conserved and if it occurs more or less frequently in the presence of one or more genes, suggesting that the genes may be either physically or functionally linked.

Alternatively, questions regarding variations over space and time can be addressed using a phylogenetic approach (see also Chapter 7). Phylogenetics reconstructs the ancestral relationships between groups, which can be a particular gene, sets of genes, microbial strains, or species. Phylogeny looks at current relationships and guesses what is the most recent common ancestor. Presumably all organisms, if we went back far enough, have a common ancestor; phylogeny gives our best guess of how the groups evolved from that ancestor. Depending on the information available, phylogeny shows which group evolved most recently, the genetic distance between the groups, and the evolutionary time since the groups diverged. When we have information about the time of isolation, for example, if a rapidly evolving organism has been isolated multiple times from the same individual over time, our guesses as to evolution can be greatly improved. Over a relatively short time period, some microbes can evolve to be drug resistant within an individual. Phylogenetic analyses enable rapidly mutating microbes to be grouped for analysis. The analyses also give a theoretical basis for guessing the order of mutation. Phylogenetic analysis can be used to estimate the relative transmission rate among different populations, and potential pathways of transmission. One of the most common applications of phylogenetic analysis in epidemiology is during an outbreak. For example, the results from a pulsed-field gel electrophoresis (PFGE) or sequencing can be analyzed and displayed using a dendrogram to show how the isolates cluster. Genetic distance might be measured using difference in restriction sites, as is done for trees constructed using PFGE, nucleotide sequences in one or more genes, intergenetic sequences, or amino acid sequences.

A phylogenetic analysis of the isolates from an outbreak confirms or refutes the epidemiologic data. Are the time–space clustered isolates the same, suggesting a common source outbreak; closely related, suggesting a propagated outbreak; not related; or the result of contamination, a "pseudo-outbreak"? Pseudo-outbreaks are not uncommon in settings with high background (endemic) rates, such as hospitals, or when a policy change or new technology begins monitoring for a new organism. A cluster of skin and site swabs specimens from a single clinic did not grow bacteria normally found on the skin or in wounds but rather a *Pseudomonas*-like fermenting bacteria, subsequently identified as *Cupriavidus (Ralstonia, Wautersia) pauculus.*[4] An epidemiologic investigation determined that the swabs were moistened with tap water before specimen collection. When cultured, the tap water also contained *Cupriavidus (Ralstonia, Wautersia) pauculus.* Six outbreak isolates and the tap water isolate were subjected to PFGE and analyzed to produce a dendrogram. The tree was calculated using Pearson pairwise correlation with the average linkage method (Figure 11.2).

The ability to determine the order of transmission helps us understand the introduction of genes or strains into the population. Was there a single introduction or were there multiple introductions of a new strain or subtype? Are there single or multiple transmission chains stemming from a single source? This type of information provides insight into characteristics that increase transmission and into the transmission mode and infectivity of the microorganism. Determining transmission order also has legal applications. Phylogenetic analysis has played an important role in high-profile legal cases of suspected deliberate HIV infection. Although phylogenies can effectively demonstrate that isolates are not related, they are not as good at the reverse, as the same mutations can occur independently. This

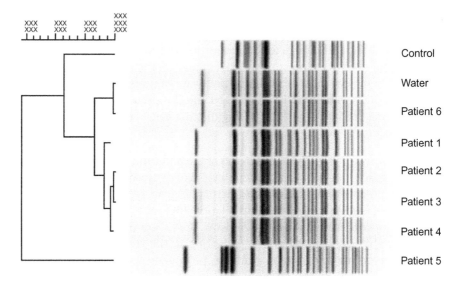

**FIGURE 11.2**

Dendrogram of PFGE of six *Cupriavidus (Ralstonia, Wautersia) pauculus* isolated from patients and one from tap water. The tree was calculated using Pearson pairwise correlation with the average linkage method. *Source: Reproduced, with permission, from Balada-Llasat et al.*[4] *(2010).*

phenomena, known as parallel or convergent evolution, is not uncommon when an organism is exposed to the same selective pressure, for example, to antiviral or antibacterial drugs. Convergent evolution has been identified in antiviral mutations of HIV.[5] As the number of sequences of microbes deposited in databases increases, our ability to determine evolutionary relationships will only improve, because we will be better able to identify mutations known to have multiple origins and take them into account during analysis.

Evolutionary relationships can also provide insight into the emergence and spread of more virulent clones associated with disease. Groups of closely related genotypes or clonal clusters emerge from a founding genotype that becomes predominant, either through selection or via genetic drift, forming clonal complexes. A clonal complex is a biologically related grouping that has diversified recently from a common ancestor.[6] Closely related strains can look quite different using a gel-based method if one strain integrates a transposon into its genome. To determine common ancestry of the strains, it is useful to limit the analysis to genes that constitute the genetic core of the particular species and compare differences only in those genes. Multilocus sequence typing (MLST) uses this strategy. In MLST, housekeeping genes are selected, usually seven, and their sequence is determined. Each individual sequence is assigned to an allele according to a common database. The allele numbers are concatenated and the resulting number is called a sequence type (ST). Phylogenetic analysis of bacteria can be complicated by recombination, but recombination rates are low in housekeeping genes over short time periods, making MLST data ideal for evolutionary analyses (see Chapter 6 for more details on MLST).

MLST systems have been developed for many bacterial pathogens, and there are public databases available. MLST enables phylogenetic analysis that can be displayed using a dendrogram. An alternative method of displaying phylogenetic analyses of MLST is eBURST. An eBURST analysis displays MLST data as clusters of related isolates based on ST. The program groups the related isolates into putative clonal complexes, and predicts the genotype that founded each clonal complex using a bootstrap method to estimate the likelihood that the assignment of the putative founding genotype is correct.[6] Figure 11.3 shows an eBURST analysis of 2001 isolates of *Campylobacter jejuni*. The diagram shows the estimated pattern of descent from the primary founder for a single sequence type, here ST 21.

The arrow points to the founding genotype. The circles are clusters of related isolates; the user sets how many alleles must be in common within each ST in the clusters. (Allele here is a reference to variations in the sequence in one of the loci.) The circle size is proportional to the number of related isolates. The smallest circles represent single locus variants. Single locus variants have a ST that is the same as the reference for all but one allele.

If nucleotide substitution for a neutral mutation occurs at a constant rate, which is true for some genes and proteins not under selective pressure, the neutral mutation rate can be used as a molecular clock to estimate the time since the population shared a common ancestor. When tied with ancillary information, such as the fossil record, or the timing of strain isolation, the time since a common ancestor can be determined more accurately. The microbial clock, however, cannot be used to estimate the absolute age of a microbial strain or species, because some microbes go through events that purge diversity (selective sweeps).[7]

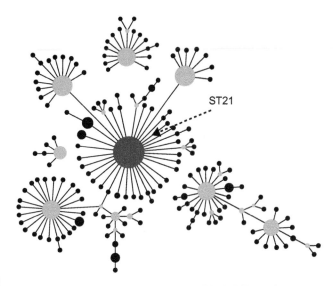

**FIGURE 11.3**
eBURST display of *C. jejuni* ST-21. The arrow points to the ancestral genotype. The size of the circles represent the number of isolates with the ST in the input data (here there were 2001 isolates). Solid lines indicate links between primary or sub founders and single locus variants.
*Source: Reproduced, with permission, from Feil et al.*[6] *(2004).*

Influenza A is an RNA virus that frequently undergoes genetic reassortment in the genes that code for neuraminidase (N) and hemagglutinin (H), two surface proteins important for influenza pathogenesis. Reassortment enables the virus to sustain transmission in the face of immunity to these proteins. Influenza A viruses are classified into serotypes based on N and H, for example, influenza A, H1N1. In 2009, a new subtype of influenza A H1N1 emerged that rapidly spread throughout the world via human contact. Understanding the origin of this subtype and previous subtypes that caused pandemics can provide important insights into how to design more effective surveillance systems. If we were better able to recognize potential new pandemic strains, there would be more lead time to implement strategies to lessen the spread of disease. A phylogenetic analysis found that the 2009 influenza A H1N1 probably originated from influenza A circulating among swine, because the genetic sequences were similar.[8] As swine influenza can be transmitted to humans, and swine have been the source of previous human epidemics, humans might be better protected by implementing ongoing surveillance for influenza among swine. A molecular clock analysis suggested that the common ancestor of the 2009 H1N1 strain emerged in mid-September of 2008.[9] Further, the 2009 H1N1 strain has gene segments from both North American and Eurasian swine influenzas, some of which are of avian origin. The origins by gene segment and years of introduction are shown in Figure 11.4.[10]

Our conceptualization of infectious disease processes and movements through populations has been heavily shaped by the Koch–Henle postulates, which define infectious disease as stemming from the presence of a single pathogen. This conceptualization has been very effective in reducing the transmission and pathogenesis of many of the major infectious scourges of humankind. However, we are now beginning to appreciate how interactions among microbes may influence disease risk. For some infections, like those that cause dental decay and infections of the gums (periodontal disease), no single pathogen has been identified, and the diseases have been designated as polymicrobial. For others, infection by one organism mediates risk of infection by another. Several viral infections depress immune function, notably HIV, but also measles and chickenpox, increasing risk of secondary bacterial or fungal infections. Some infections, like genital ulcerative diseases, cause lesions, which enhance transmission of blood-borne infections (like HIV and hepatitis B). Even more complex are diseases that seem to be infectious in nature for which no organism or group of organisms has been identified as causal. For example, inflammatory bowel or Crohn's disease is accompanied by a disrupted microbiota that may either be the cause or result

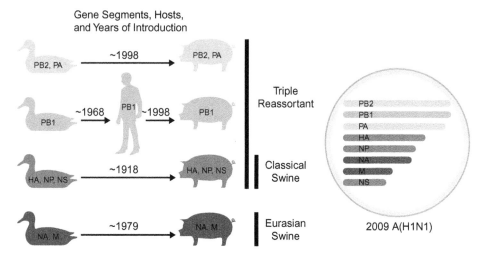

Gene Segments, Hosts, and Years of Introduction

**FIGURE 11.4**

Graphical description showing results of phylogenetic analysis of different gene segments of the 2009 A(H1N1) virus. Each of the gene segments has a different lineage; this combination of segments had not be seen anywhere previously. *PB2*, Polymerase basic 2; *PB1*, polymerase basic 1; *PA*, polymerase acidic; *HA*, hemagglutinin; *NP*, nucleoprotein; *NA*, neuraminidase; *M*, matrix gene; *NS*, nonstructural gene. *Source: Reproduced, with permission, from Garten et al.*[10] *2009.*

of the disease. To understand polymicrobial disease or diseases of disrupted microbiota, our analysis must focus on the microbial community. Describing microbial communities based on the genomes present, rather than by culturing each organism separately, is called metagenomics. Some questions we might ask about the microbial community are shown in Table 11.4.

Analyses of interactions among microbial species draw on ecological theory, which uses analytical techniques to understand how microbial communities are assembled and how they change over time. These analyses use measures of community structure, such as indices of richness, diversity, and evenness. Richness refers to the number of individuals in a sample, diversity to the number of different species observed, and evenness to the relative abundance of the observed species (see Chapter 7). Dethlefsen and colleagues[11] compared the richness, diversity, and evenness of intestinal microbiota among three healthy individuals before, during, and after treatment with ciprofloxacin, a broad-spectrum antibiotic (Figure 11.5). They used pyrosequencing of two variable regions of the 16S rRNA

**TABLE 11.4 Analytical Questions at the Molecular Level about the Microbial Community**

- What is the composition of the microbial community? (What species are members of the community?)
- What is the richness of the microbial community? (How many species are there?)
- What is the structure of the microbial community? (What are the relative abundances of the microbes present?)
- How variable is the microbial community structure within an individual over time?
- Does the composition and/or structure of the microbial community change with spatial distance? Is there geographic variation?
- Does the composition and/or structure of the microbial community change with time? Is there seasonal variation?
- Is the composition of the microbial community associated with a specific outcome or exposure?

gene to identify the number and abundance of different operational taxonomic units (OTUs) in the samples, which was unit of analysis. In this study, most of the OTUs could be resolved into different genera. Multiple samples were tested for the different time points. Richness, diversity, and evenness were lowest when the participants were undergoing antibiotic therapy; although the microbiota in all participants returned to the pretreatment levels after therapy, there were some taxa that failed to recover. This suggests that the community is resilient to disruption, and there is some redundancy in the function(s) of the inhabiting microbiota.

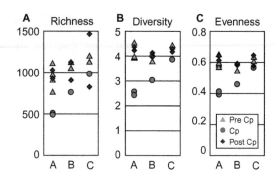

**FIGURE 11.5**

Comparison of richness, diversity, and evenness of intestinal microbiota among three healthy individuals (A, B, and C, notes on the x-axis), before, during, and following treatment with ciprofloxacin. *Source: Reproduced, with permission, from Dethlefsen et al.[11] (2008).*

## 11.7 INTERACTIONS OF MICROBES AND HOST

Because epidemiology is concerned with health and disease, most molecular epidemiologic research focuses on the interaction of a (usually single) microbe with its host. Typical questions relating to these interactions are shown in Table 11.5, with the first two focusing on microbial characteristics associated with disease and host or microbial biomarkers of disease or prognosis. Analyses of these questions can be framed similarly to a typical epidemiologic categorical data analysis, using odds, risk, or rate ratios as the measure of effect depending on the study design. The presence of a particular microbial characteristic can be related to a measure of pathogenesis, persistence, or transmission in a $2 \times N$ table for dichotomous measures. Analysis of variance for continuous measures or a multivariate model can also be used. This type of analysis treats the microbial characteristic just like any other exposure variable – a causal factor that can be independent of or interact with other exposure variables leading to the outcome of interest. Uropathogenic *E. coli* have certain characteristics that enable them to inhabit and persist in the urinary tract. One such factor, hemolysin, was hypothesized to be associated with risk of recurring urinary tract infection (UTI). In a prospective cohort study of women with first UTI caused by *E. coli*, however, when the first UTI carried hemolysin, the risk of a second UTI was significantly decreased.[12]

Microbial measures or biomarkers may arrive from the laboratory in various formats that are more or less amenable to analysis. The laboratory data may be dichotomous (present/absent), which poses no difficulties. It is more likely, however, that the interpretation and transformation into a usable measure is more complex. For example, there may be a range of genetic variants with different SNPs or indels occurring at various frequencies either singly or together. Variants may or may not result in a change in protein structure depending on whether they are in coding regions or if nucleotide changes are synonymous. Therefore, the first task of analysis is to determine what is a meaningful coding of the observed variation for analysis. This decision should be made on biological grounds (such as a phylogenetic analysis) to avoid choosing groupings based on the statistical significance of the resulting association with the outcome of interest. The resulting groupings can be used in further epidemiologic analyses, such as cross-tabulations with the outcome or exposure of interest.

**177**

**TABLE 11.5** Analytical Questions Relating to Interaction of Microbe with the Host (Molecular with Individual Scale)

- Do specific microbial characteristics occur more frequently during diseased than healthy states? Is a microbial characteristic associated with pathogenesis, persistence, or transmission?
- What biomarkers (either from the host or microbe) predict disease? Disease prognosis?
- Are there microbial community structures that correspond to health or disease?

**TABLE 11.6** Analytical Questions Relating to Interactions of Individual Humans (Human Individual Scale)

- What was the direction of transmission (who infected whom)?
- What factors distinguish those individuals most susceptible to colonization? To disease?
- What factors distinguish those individuals most likely to transmit infection to another?
- What is the basic reproduction number ($R_0$)?
- What is the probability of transmission per infected contact?
- How long is an individual infectious?
- Following recovery, how long is an individual immune?
- Can an individual be infected with different strains of the same microbe (superinfected)? Reinfected with the same strain?

Metagenomics has made it considerably easier to capture what diverse microbial populations might be present in a specific niche. This makes it possible to assess how the microbial community interacts with an invading pathogen; to observe the resistance and resilience of microbial communities to disruption, be it a pathogen or exposure to a chemical or therapeutic agent; and to identify diseases of disrupted microbiota. To analytically capture community structure requires using measures from ecology. These parameters include measures of the diversity of species present, their relative abundance, and the similarity of species between individuals. Once estimated, these measures can be used as predictors in standard statistical models.

## 11.8 HUMAN–HUMAN INTERACTIONS

Investigations of human–human interactions uses the human host as the unit of analysis; the outcome is whether the microbe is present (Table 11.6). Depending on the microbe of interest, determining the direction of transmission can be done based on a combination of host characteristics, such as when symptoms first appeared and if the infecting microbe typed identical between hosts, or for a rapidly mutating microbe, based on the types of observed changes. Analyses determining susceptibility to colonization and disease must have some measures of exposure to the microbe of interest and (ideally) infectious dose received as well as host immune status. These are difficult parameters to measure accurately outside a laboratory, but if available, the analysis is fairly straightforward using standard epidemiologic strategies. Identifying individuals most likely to transmit infection requires measures of host behavior and of the microbial characteristics, because both influence transmission. The immune state of those exposed is also needed if it is available. The analysis is again relatively straightforward using standard epidemiologic strategies.

The reproductive number ($R_0$) is a measure of the overall infectivity of the microbe: the average number of new infections that arise from each case in a totally susceptible population. Estimates of $R_0$ are extremely important for predicting the rise and fall of an epidemic. $R_0$ can be estimated directly in natural experiments when a new microbe is introduced into a wholly susceptible population using contact tracing. The contacts of all cases are tested, and the average number of secondary cases per diagnosed individual is computed. $R_0$ also can be estimated by the ratio of life expectancy to age at infection, the rate at which an epidemic is growing, or from the proportion remaining susceptible in a population after the epidemic has subsided.[13] It is also possible to estimate analytically if the components of $R_0$ are known. $R_0$ is equal to the product of the transmission probability per infectious contact, duration of infectiousness, and rate of contact. Like estimating $R_0$ directly, estimating the components of $R_0$ is more a study design than analytical problem and requires conducting prospective studies that take appropriate measurements. Lastly, $R_0$ can be estimated directly from sequence data.[14,15] In 2010, I only found estimates based on sequence data for viruses.

Several methods were used to estimate $R_0$ for the 2009 H1N1 influenza epidemic. Using data collected in Mexico at the start of the epidemic, Fraser and associates[16] first estimated $R_0$ by assuming that the increases in the number of cases was exponential, and that the generation time was similar to that observed for other influenza strains. Second, they used a Bayesian coalescent population genetic analysis. Third, they fit an epidemic model to outbreak data, and lastly, they determined a time-dependent estimate based on time series reported data of confirmed cases. The estimates ranged from 1.31 to 1.97 depending on the method.

The need to estimate $R_0$ and its components is often underappreciated by those conducting studies. Accurate assessments of these parameters are critical for predicting spread of disease and endemicity, and determining the proportion of the population that needs to be immune to stop spread (herd immunity), which are needed for planning vaccination strategies. Mathematical models used in making predictions are often hamstrung by the limited data available in the literature; even when these parameters might be estimated from the data, they may not be reported. Therefore it is important that molecular epidemiologic studies measure and report either $R_0$ or its components.

## 11.9 INTERACTIONS OF HUMAN POPULATIONS

Microbes causing human disease are not randomly distributed among human populations because they follow specific transmission pathways. Some subpopulations may, for a variety of reasons, act as either sources or reservoirs of infection. Discovering why this is so is the motivation behind the key analytical questions relating to interactions of human populations (Table 11.7).

Although epidemiologists often classify infections by transmission mode, fecal–oral, respiratory, vector-borne, and so on, most infections spread via multiple transmission modes. HIV, for example, is spread by sexual contact, via blood and blood products, and from mother to child; polio can be transmitted via water and the respiratory route. Group B *Streptococcus* is transmitted by sexual activity, from mother to child, and probably via food. Usually, however, one transmission mode acts strongly during an outbreak or epidemic, which makes analysis of disease outbreaks particularly useful for detecting new transmission modes. Outbreaks are not always detected or reported. There are many infections whose endemic rates are unacceptably high, such as *Staphylococcus aureus* infections in hospitals, which have ongoing high rates of transmission, obscuring outbreaks. *S. aureus* is presumed to spread in hospitals primarily by person-to-person direct contact, with hospital personnel acting as vectors, but both respiratory spread and spread via fomites occur. Determining the relative contributions of the different transmission modes is best estimated using a mathematical simulation model informed by basic parameters derived from a well-conducted study.

Infections are not randomly distributed in a population; usually there are subgroups with high transmission rates and subgroups with low transmission rates. Gonorrhea, for example, has a fairly narrow geographic distribution, whereas herpes, also sexually transmitted, is more broadly distributed. Identifying groups with high transmission rates, "core groups," and the reasons for that high transmission is essential for planning disease

**TABLE 11.7** Analytical Questions Relating to Interactions of Human Populations (Human Population Scale)

- What are the relative contributions of each transmission pathway to infection flow?
- Which populations act as sources or reservoirs of infection?
- What are the most important factors for determining infection spread following introduction from a reservoir?

control and prevention strategies. Analyses that describe the spatial relationships, for example, geographic mapping of the distribution of different microbial strains or ecological measures of spatial distance by strain, are useful strategies for identifying reservoirs. Taking advantage of molecular typing methods in such analyses can distinguish between different transmission systems and whether they overlap. Further identifying local hot spots for transmission can enhance screening and treatment programs. This was demonstrated in a tuberculosis screening program. The Tarrant County Health Department implemented a tuberculosis screening program based on the location of clusters of isolates with the same genotype. The yield of this geographically targeted program was significantly higher than the screening program based on risk factors; furthermore, ongoing transmission was reduced.[17]

An entire branch of infectious disease epidemiology has evolved around the emergence and re-emergence of infections. Microbes evolve at a much faster time scale than humans; the adaptation of zoonoses to humans, the evolution of human pathogens to increased virulence or resistance to existing treatments, and the adaptation to new transmission modes represent an exciting area of research. A post hoc examination of an outbreak of an emerging or re-emerging infection often leaves the impression that the outbreak was the result of a single introduction. However, careful molecular evidence, in the form of serology studies and microbial typing, suggests this is not generally the case; usually there are multiple introductions. One or several introductions may lead to sufficient transmission chains to cause an outbreak. By studying these situations and determining the difference between successful and unsuccessful introductions, we can better predict and prevent future outbreaks. Phylogenetic analysis suggests that methicillin-resistant *S. aureus* (MRSA) is not a single entity but the result of multiple importations of the genes coding for resistance into local circulating strains,[18] with the dissemination in resistance primarily local rather than global.

## References

1. A.G. Dean, K.M. Sullivan, M. Soe, OpenEpi: Open Source Epidemiologic Statistics for Public Health (website), Version 2.3, Available at: http://www.openepi.com (accessed 08.08.10).

2. R.J.A. Little, D.B. Rubin, Statistical Analysis with Missing Data, Wiley, Hoboken, 2002.

3. E.A. Archie, G. Luikart, V.O. Ezenwa, Infecting epidemiology with genetics: A new frontier in disease ecology, Trends Ecol. Evol. 24 (1) (2009) 21–30.

4. J.M. Balada-Llasat, C. Elkins, L. Swyers, et al., Pseudo-outbreak of *Cupriavidus pauculus* at an outpatient clinic related to rinsing culturette swabs in tap water, J. Clin. Microbiol. 48 (7) (2010 Jul) 2645–2647.

5. D. Pillay, A. Rambaut, A.M. Geretti, et al., HIV phylogenetics, BMJ 335 (7618) (2007) 460–461.

6. E.J. Feil, B.C. Li, D.M. Aanensen, et al., eBURST: inferring patterns of evolutionary descent among clusters of related bacterial genotypes from multilocus sequence typing data, J. Bacteriol. 186 (5) (2004) 1518–1530.

7. E.C. Holmes, Evolutionary history and phylogeography of human viruses, Annu. Rev. Microbiol. 62 (2008) 307–328.

8. G.J. Smith, D. Vijaykrishna, J. Bahl, et al., Origins and evolutionary genomics of the 2009 swine-origin H1N1 influenza A epidemic, Nature 459 (7250) (2009) 1122–1125.

9. Rambaut A., Epidemiology and molecular clock analysis (website), Available at: http://tree.bio.ed.ac.uk/groups/influenza/wiki/4b60e/Epidemiology_and_molecular_clock_analysis.html, 2009 (Updated 10.05.09; accessed 08.08.10).

10. R.J. Garten, C.T. Davis, C.A. Russell, et al., Antigenic and genetic characteristics of swine-origin 2009 A(H1N1) influenza viruses circulating in humans, Science 325 (5937) (2009) 197–201.

11. L. Dethlefsen, S. Huse, M.L. Sogin, et al., The pervasive effects of an antibiotic on the human gut microbiota, as revealed by deep 16S rRNA sequencing, PLoS Biol. 6 (11) (2008) e280.

12. B. Foxman, L. Zhang, P. Tallman, et al., Virulence characteristics of *Escherichia coli* causing first urinary tract infection predict risk of second infection, J. Infect. Dis. 172 (6) (1995) 1536–1541.

13. K. Dietz, The estimation of the basic reproduction number for infectious diseases, Stat. Methods Med. Res. 2 (1) (1993) 23–41.

14. E.M. Volz, S.L. Kosakovsky Pond, M.J. Ward, et al., Phylodynamics of infectious disease epidemics, Genetics 183 (4) (2009) 1421–1430.

15. O.G. Pybus, M.A. Charleston, S. Gupta, et al., The epidemic behavior of the hepatitis C virus, Science 292 (5525) (2001) 2323–2325.

16. C. Fraser, C.A. Donnelly, S. Cauchemez, et al., Pandemic potential of a strain of influenza A (H1N1): Early findings, Science 324 (5934) (2009) 1557–1561.

17. P.K. Moonan, J. Oppong, B. Sahbazian, et al., What is the outcome of targeted tuberculosis screening based on universal genotyping and location? Am. J. Respir. Crit. Care Med. 174 (5) (2006) 599–604.

18. U. Nubel, P. Roumagnac, M. Feldkamp, et al., Frequent emergence and limited geographic dispersal of methicillin-resistant *Staphylococcus aureus*, Proc. Natl. Acad. Sci. U.S.A. 105 (37) (2008) 14130–14135.

# CHAPTER 12

# Human and Animal Subject Protection, Biorepositories, Biosafety Considerations, and Professional Ethics

Researchers are morally obligated to conduct research in an ethical fashion, and there are many rules and regulations in place to assist them in doing so. It is the responsibility of the principal investigator of a research project to ensure the health and safety of study personnel, to conduct the research project in an ethical manner, to make sure that materials are handled and stored safely, and to honor the implied obligations of a researcher to society, that is, to further scientific knowledge, disseminate the knowledge gained, and to do no harm.

During the design, conduct, and analysis of research on humans and animals an investigator is faced with several moral dilemmas. Given the current state of knowledge and available methodologies, is it morally justified to ask human beings to participate in the proposed research? Are the anticipated results worth the associated risks to participants and study personnel? Does the study protocol respect an individual's rights? Molecular epidemiologists face these and additional moral dilemmas associated with conducting public health research and using stored specimens. One such dilemma is a conflict between respect for a person's autonomy and the moral obligation to improve the public's health. To protect the public from disease may require interfering with individual rights, such as requiring someone with tuberculosis to be treated or quarantining someone who is actively shedding *Mycobacterium leprae*, which causes Hansen disease (also known as leprosy). To identify these individuals requires active surveillance of the population; the resulting data can be an important resource for research projects. If the data are collected for one purpose (public health surveillance) and used for another (research), what is the responsibility of the investigator to the participant? Molecular tools enable secondary analyses of existing specimen collections in biorepositories. Is it ever ethical to conduct studies on these samples for which the persons contributing samples did not specifically consent? Identifying and responding ethically to these moral dilemmas is the purview of the field of bioethics.[1] In this chapter, the intention is to acquaint the reader with the rules and regulations in place intended to assist the researcher in behaving ethically during the design, conduct, and analysis of a study. For an in-depth presentation of bioethics, the reader is referred to the extensive literature on the topic.

## 12.1 WHAT IS RESEARCH?

Research is subject to rules and regulations that medical treatment and public health activities are not. Whether or not medical treatment and public health activities are research does not change the investigator's obligation to behave in an ethical fashion,

Molecular Tools and Infectious Disease Epidemiology.
© 2012 Elsevier Inc. All rights reserved.

but it does determine the procedures that must be followed. Thus it is important to know what constitutes "research" in order to comply with both the spirit and letter of existing rules and regulations. The use of a medical therapy or diagnostics in an individual patient does not constitute research, but tabulating results across a clinical population to identify relationships is research. Similarly, surveillance and outbreak investigations are public health activities, but if a new infectious agent is discovered or a novel transmission system for a known pathogen is identified, the results may be considered research. In both medicine and public health, when an activity crosses the line to become research is often fuzzy.

The National Institutes of Health defines research as: "A systematic, intensive study intended to increase knowledge or understanding of the subject studied, a systematic study specifically directed toward applying new knowledge to meet a recognized need, or a systematic application of knowledge to the production of useful materials, devices, and systems or methods, including design, development, and improvement of prototypes and new processes to meet specific requirements."[2] Key terms in the definition are that the activity is systematic and that it involves generating or applying new knowledge. Medical treatment is not intended to generate or apply new knowledge but to make a particular patient well. However, an astute physician may make an observation that is new and wish to report that observation. The rules and regulations are intended to ensure that when a patient becomes a subject of investigation rather than a subject of medical care that the physician continues to behave ethically, adhering to the principles of nonmaleficence, doing no harm, and beneficence, maximizing benefits over risks.

The patient (usually) voluntarily enters into the relationship with a physician, and consent is implied (although consent may be obtained for specific therapies). The physician partners with the patient to make the patient well: the physician recommending treatments and the patient following them. Ideally, the physician and patient work together to find the optimal individualized treatment strategy. However, in this voluntary arrangement the partnership is not equal. The physician has knowledge the patient does not, and the patient may be concerned that questioning or refusing his or her physician's advice may lead to poorer care. Physicians are people too, and their beliefs regarding the value of a particular treatment, or the opportunity to recruit patients to participate in drug treatment trial, or eagerness to try out a new therapy may make them less sensitive to a particular patient's needs. In medical research, where the intention is to evaluate a new test or therapy, the physician/investigator is focused on the evaluation process rather than the individual. These evaluations require sufficient numbers of individuals so that relatively rare events can be detected, and that the physician/investigator can be certain that apparent treatment effectiveness is real and not due to chance. This can make the physician/investigator more likely to pressure a patient that is a particularly good candidate to take a therapy, and to be more reluctant to change or modify the therapy when required. It is this difference in perception of the patient – as a subject of investigation rather than a patient to make well – that might lead a physician/investigator to behave less ethically. This difference in perception is what rules and regulations protecting human subjects are intended to correct.

Public health research has been defined as "a systematic investigation undertaken to develop or contribute to generalizable knowledge as opposed to public health activities which are undertaken primarily to benefit the individual by preventing disease or injury or to improve the health of the community."[3] Using intention to develop or contribute to generalizable knowledge as the litmus test for implementing ethical safeguards puts the investigator in a difficult situation. An outbreak investigation is a public health activity. The results do not affect those already ill, but can be generalized to protect the health of others. Consider a food-borne outbreak that identifies a novel vehicle for transmission, such as hepatitis A in strawberries or salmonella in peanut butter. Should the information not be published in the scientific literature because the study that identified it was part of public health activity

184

---

**TABLE 12.1 Some Allowable Exemptions to Institutional Review of Human Research**

- Research relating to the development and evaluation of educational practices
- Research involving educational testing, unless participants can be identified and the results might put them at risk of criminal or civil liability or damage their financial standing, employability, or reputation
- Research on publicly available existing data, documents, records, pathological specimens, or diagnostic specimens, or if the data are anonymous

---

and not research, and so not subject to ethical review? At the time of this writing, the decision as to which or whether public health investigations are subject to ethical review and require informed consent is in the hands of the individual local, state, and federal agencies. However, human subjects review is probably necessary for publication in peer reviewed journals.

Because the line between the practice of medicine and research or between public health activities and research can be fuzzy, it can be difficult to determine if a project is research. A major problem encountered by institutional review boards is investigators who either did not know that the protocol they intended to follow required institutional review or assumed it was exempt and did not require review. Therefore, for the individual investigator working in a medical or public health setting, it is best to err on the side of caution and submit all study protocols to the appropriate reviewing body. Some research can be deemed exempt (Table 12.1), or the need to obtain informed consent can be waived, but the determination of exemption or waiver of consent should be made by the reviewing body. Even if an exemption or waiver of consent is given, the reviewing body weighs the risks and benefits of the proposed project. Often molecular epidemiologic studies are deemed exempt because they use existing data, specimens, or diagnostic specimens, and the links to identifying information have been destroyed. However, any time an investigator uses data collected from human subjects, the investigator should apply for approval from the appropriate committee. By subjecting his or her proposed protocol to external review, the investigator obtains external validation that the proposed research meets ethical standards. It also protects the investigator, and his or her peers at their home institution; violations to one human subject protocol have led to the shutdown of all ongoing research at the institution.[4]

## 12.2 WHY RESEARCHERS ARE OBLIGATED TO BEHAVE ETHICALLY

While each individual should feel morally obligated to behave ethically, for example, not to steal, a researcher is additionally obligated to behave in a moral and ethical way because of the trust the research participants and study personnel place in him or her. Research should further knowledge and should benefit people, if not the individual participant. The rights of participants should be protected, and any risks should be outweighed by the potential benefits. The researcher is further obligated to behave ethically because of the trust society has shown in him or her, in supporting the research effort. The funder trusts that the researcher will conduct research in an ethical fashion, from protecting study participants to reporting the results truthfully in the scientific literature. We have locks on our doors to keep honest people honest; similarly there are a number of rules, regulations, and oversight committees to help researchers behave ethically. These regulations are relatively new; institutional review boards were implemented in 1974.[5]

The impetus for government oversight was the many abuses that have occurred under the guise of research. Though one might argue that acquisition of knowledge is always a "good," there are many paths to acquiring that knowledge, and, as noted in the quote by Hans

Jonas (below), the benefits of a speedy yet immoral path are completely outweighed by the accompanying erosion of moral values.

> Let us not forget that progress is an optional goal, not an uncompromising commitment. A slower progress in the conquest of disease would not threaten society, grievous as it is to those who have to deplore that their particular disease be not yet conquered, but that society would indeed be threatened by the erosion of those moral values whose loss possibly caused by too ruthless a pursuit of scientific progress, would make its most dazzling triumphs not worth having.
>
> **Hans Jonas,[6] 1976**

The potentially most flagrant and well-known abuses conducted under the pretense of research occurred in Nazi Germany and in Japan during World War II. This "research" included freezing, infecting individuals with diseases in order to test the effectiveness of alternative therapies, and exposing persons to chemicals to test toxicity – all without their consent. Similar, but perhaps less well documented, events have occurred worldwide. In the United States the Tuskegee syphilis experiment was conducted by the U.S. Public Health Service to describe the natural history of syphilis. This study regularly examined participating individuals as they acquired syphilis, documented the myriad of physical conditions that the disease causes, and never offered any therapy – indeed, actively prevented treatment – long after effective therapies were available.[7–9] However, there have been many other incidents; examples from a range of medical settings were documented in the landmark article by Henry K. Beecher,[10] and such incidents continue. For example, in 2008, researchers at the Department of Veterans Affairs (VA) Central Arkansas Health Care System (CAVHCS) performed HIV tests without consent, and may not have obtained consent from some participants or proper consent from demented patients.[11] Therefore there are very sound reasons for requiring government oversight for all experiments using humans, particularly when there is no direct benefit of the research for the participating patient. Because animals are incapable of giving their consent, and much animal research requires sacrificing the animal to determine study results, the requirements are even more stringent.

## 12.3 PROTECTION OF HUMAN SUBJECTS

By legal requirement and standards of research ethics all studies involving humans in the United States and other nations must be reviewed to determine whether the rights and welfare of study participants are protected during the research process (*United Nations Educational, Scientific and Cultural Organization* [UNESCO] adapted the universal declaration on bioethics and human rights in 2005). The reviewing body, called the Institutional Review Committee in the United States, determines whether the research protocol is consistent with the ethical principles of respect for autonomy, nonmaleficence, beneficence, and justice (Table 12.2).

There is extensive literature on these principles (see for example *Principles of Biomedical Ethics* by Beauchamp and Childress[12]), which will be briefly summarized here.

### TABLE 12.2 Ethical Principles Behind Informed Consent Process

- Respect for autonomy: protects the dignity and autonomy of individuals, and requires that people with diminished autonomy be provided special protection
- Nonmaleficence: do no harm
- Beneficence: protect individuals by maximizing anticipated benefits and minimizing possible harms
- Justice: treat individuals fairly

Respect for autonomy means that the researcher should respect the right of the individual to make his or her own decisions. For consent to be voluntary, any incentives given should be of token value in order to not be coercive. For example, a potential participant who is in jail cannot be promised parole review as an incentive. Similarly, there should be no penalty for not participating. Potential participants should not be concerned that the decision to participate will influence the relationship with their physicians or the quality of medical care received. If the potential participant cannot make an informed decision because of illness, mental impairment, or youth, a proxy must be selected. There should be protections in place to ensure that the proxy acts in the potential participant's best interest. Respect for autonomy is operationalized by obtaining informed consent. The informed consent document is intended to provide the potential participant with enough information to make an informed decision. The document should include the purpose of the study, what the participant is being asked to do, and any risk or benefits associated with participation (Table 12.3). The consent form also explicitly states that the participant can refuse to participate or withdraw at anytime without penalty. Some institutions have templates and require specific language in specific order; others do not. The need for a written informed consent can be waived, but the study protocol should ensure that the study is conducted in a way that respects the autonomy of participants.

Nonmaleficence means to do no harm. Nonmaleficence is part of the Hippocratic oath taken by physicians. In medicine, nonmaleficence means that a physician will provide only treatments she or he believes to be effective and not intentionally harm a patient – beyond the bounds of what is required for therapy. Surgery, for example, involves doing harm to do good. In public health, nonmaleficence also requires balancing competing concerns. Ordering a food implicated in a food-borne outbreak withdrawn from the market with the intention of preventing future illness does harm to the manufacturer or distributor and the businesses distributing the food. In a research project, there are multiple opportunities for a well-meaning investigator to do harm. Collecting biological specimens puts the participant and research personnel at risk of harm; the extent of harm depends on the specimen. Collecting blood causes some pain and the potential of a hematoma. Handling blood has a risk of acquiring disease. The information discovered by the investigator, such as increased risk of disease, may cause undue pain and suffering. If illegal behaviors are identified, such as substance abuse, loss of confidentiality may jeopardize the participant's livelihood and put him or her at risk of prosecution.

Conducting a study in a community, regardless of the stated purpose, is effectively an intervention. Asking for a dietary, sexual, or medical history forces respondents to examine their behavior; collecting a blood or urine or fecal specimen can cause anxiety about the results of the specimen testing. Further, study participants may be looking for validation of specific behaviors, which may include ones under study as potential risk factors. Thus study personnel should be carefully trained and study materials thoughtfully prepared so

**TABLE 12.3 Elements Generally Required in an Informed Consent Document**

- A description of the project and its purpose, and a clear statement that it is research
- What the participant is being asked to do, whether the procedures are experimental, whether alternative treatments might be advantageous, and what the risks are to the participant
- What benefits to the participant or to others are expected from the research
- If the data will be confidential, and how confidentiality will be maintained
- Whether there will be any compensation for participation, and if medical treatment will be provided should injury occur
- Whom to contact with questions about the research, the participant's rights, or in the case of research-related injury

they do not inadvertently give misinformation, create anxiety, or cause stigma. However, the investigator is morally obligated to give participants their results in a way they can understand. Sending all study participants a summary of study results creates goodwill for all researchers and gives participants a tangible benefit of participation. This can be done without compromising confidentiality: participants might address an envelope during enrollment which can be used to send study results. The envelope should be filed separately from all other study materials. If the participant wishes and consents, the results of individual medical tests can be sent to the participant's treating physician.

The need to balance the risks associated with conducting research against the potential benefits is embedded in the principle of beneficence, which means "doing good." The benefits from clinical studies range from evaluating the effectiveness of a new therapy to increasing our understanding of the underlying biology that leads to disease. In public health the benefits range from evaluating the effects of a population intervention to describing the distribution of factors relating to health and disease. The risks can be to individual participants or to society. For example, vaccination against a sexually transmitted disease may result in an increase in unplanned pregnancies.[13] Participating in a clinical trial of a new drug puts the participant at risk of unknown adverse effects related to the new drug and the known risks of either remaining untreated (if placed in placebo group) or receiving current care (if randomized to standard care). These risks should to be balanced against the benefit of identifying a better or alternative therapy. During the course of a study, the investigator might identify information that suggests there is a risk of potential harm to the health and safety of a particular study participant. Indeed, a population might be selected for study *because* they are exposed to a factor that might cause harm. The investigator is obligated to inform the study participants if they may be at risk, and similarly inform participants if a source of potential harm is detected during the course of the study, if the benefits of doing so are unambiguous. Clearly, it would be beneficial to notify individuals not to eat a certain food if it was implicated in a food-borne outbreak. Notifying participants that they have an increased risk of a disease for which early detection and treatment has no effect on disease course is of dubious benefit and may cause undue harm.

Justice means that all individuals are treated fairly. Excluding certain individuals from study can mean that the results will not pertain to them. For many years, drug trials were not conducted on women to avoid the potential harm to an unborn fetus. The result was that the proper dosages for treating women with these drugs were not based on clinical trials but by extrapolation from the results found among men. Many studies have subsequently demonstrated that there can be gender differences in drug effectiveness, as well as differences related to age and ethnic group. For the individual investigator, the principle of justice translates into making sure that the inclusion and exclusion criteria are scientifically justified. Justice also dictates that if the protocol is not ethical for you and your family, it isn't ethical for others. Similarly, if a protocol is not ethical in your home country, it isn't ethical elsewhere.

The mechanics of obtaining approval for conducting research are generally straightforward: in the United States each institution has a committee, known as the Institutional Review Board (IRB) or Human Subjects Committee, which reviews all study protocols that will be conducted under the auspices of that institution. In the United States, the IRB must by law include at least five members. The membership should be diverse with respect to professional training and gender and include the appropriate expertise to review the research under consideration at the institution. (The reader is directed to Title 45 CFR 46 for the U.S. code of regulations, accessible at http://ohsr.od.nih.gov/guidelines/index.html.[14]) Although IRBs are guided by legislation, how procedures are operationalized is particular to an institution. For example, some IRBs require that consent forms be structured according to their guidelines and others do not. Nonetheless, all IRBs review study protocols and informed consent documents with the intent of protecting the rights and welfare of study participants. IRBs can require changes in study protocols and procedures and consent forms. This often makes investigations conducted at multiple institutions challenging, because currently each IRB makes a separate determination – which may or may not agree with IRBs at other institutions. The investigator must balance the need for IRB approval at a particular institution with the research need for uniformity across research sites.

Some populations are considered particularly vulnerable, and there are additional regulations regarding their study. These populations include pregnant women, fetuses, neonates, children, and prisoners. The primary concern is that these populations may be unable to voluntarily give consent and weigh the benefits and risks of participation in research. For pregnant women, fetuses, and neonates the investigator must be able to demonstrate that the research will have "a reasonable opportunity to further the understanding, prevention, or alleviation of a serious problem affecting health or welfare."[14] To protect prisoners, the majority of the IRB reviewing the protocol must have no relationship with the prison, and a prisoner must be a member of the review. In addition, the investigator must ensure the selection for participation is fair, and participation (or nonparticipation) will **not** be taken into account by authorities, such as the parole board.

Since 1996, researchers have been required to respect the Health Insurance Portability and Accountability Act (HIPAA) in developing their study protocols.[15] HIPAA is intended to protect the privacy of individual health information. Respecting the intent of HIPAA requires changes in study protocols and additional oversight; in many institutions, the IRB is responsible for ensuring that HIPAA rules are implemented. At the simplest level, HIPAA requires that all individuals handling health-related information be HIPAA certified. It also constrains the type of information that can be released without specific informed consent. Investigators should ensure that their study protocol and study personnel are HIPAA compliant.

Some states have local laws or regulations that provide additional protection; these supersede federal regulations. These state laws or regulations address the privacy of health and medical information, genetic information, and human subjects' protection. While many of these states may have exceptions for research, not all do so, and the conditions for the exceptions vary by state.[16] Thus researchers must inform themselves of the content of any additional laws and regulations that may affect their studies.

## 12.4 STUDIES USING PREVIOUSLY COLLECTED DATA

Molecular tools enable new analyses of specimens from existing collections. For example, a throat swab may be collected and the swab frozen. Specimens may be stored in duplicate, and the extras saved in a biorepository (see Section 12.5). Although many microbes may no longer be viable, amplifying DNA coding for ribosomal genes will give a picture of the microbial population that was present. Similarly, biological material may be scraped from slides made from vaginal smears, and the microbial population determined using the same techniques. It also is possible to amplify human DNA from specimens like saliva or urine or any specimens where human cells might be present. This provides a tremendous opportunity to give added value to completed studies with stored specimens, and future value for ongoing studies, because it seems likely that new ways to explore existing specimens will be developed. These capabilities are mentioned again in this chapter because of the implications for the protection of human subjects. If the information is already collected, and the data are recorded in a way that the data cannot be linked by to the individual subjects, these data are considered exempt under the federal common rule. However, many states may have additional requirements.

Access to tissue specimens and associated data, and requirements regarding protection can vary substantially from state to state. For a multistate study, data collection in a specific state must follow the laws and regulation regarding confidentiality of medical information and informed consent for that state. If the data are sent to a central repository or laboratory for processing, the investigators must determine whether the informed consent document must specifically allow for disclosure of medical information to the repository or laboratory.[16] The lack of consistency across states poses significant challenges for multisite studies and use of data and specimens from central repositories, because the requirements of the state where data were collected must be met.

In 2004, the National Cancer Institute conducted a 50-state survey of the regulations,[16] which concluded the following:

> While few states specifically address research uses of tissue samples, numerous laws that regulate medical information, genetic testing, and the conduct of human subjects research affect the ability of researchers to collect, store, access, distribute, or analyze tissue samples and associated data. A review reveals that most of these laws focus on protecting individual privacy and confidentiality to minimize risks of harm. Some potential risks to subjects whose specimens and data are used in research could result from releases of private medical information. These risks might include loss of insurance or loss of employment. Other risks that have been identified are not necessarily related to inappropriate releases of information, such as loss of dignity or autonomy, or infringement on privacy.

> The laws in various states differ widely in scope, definitions used, and applications for research. The rules also differ depending on whether the source is public health statutes, insurance statutes, privacy laws, or laws protecting research subjects. As a result, the application of many statutes to research uses of tissue and data is unclear, with vague and conflicting definitions leading to variable interpretations and implementation.[16]

## 12.5 ETHICAL ISSUES ASSOCIATED WITH BIOREPOSITORIES

A biorepository is "a collection of human specimens and associated data for research purposes, the physical structure where the collection is stored, and all relevant processes and policies."[17] Biorepositories pose specific ethical issues (Table 12.4).[18] The participant donating a specimen, be it blood, urine, or tissue, expects that the specimen will be valued, that it will be handled and stored appropriately, and distributed as needed so it will contribute to scientific knowledge. Further, the participant expects that the users of the donated specimens will respect his or her privacy, and that the specimen(s) will be used only in ways consistent with the consent granted. As biorepositories are generally intended to extend beyond any single grant period, establishing a biorespository places considerable responsibility on the investigator and the hosting institution.

### 12.5.1 Responsible Custodianship

Specimens are precious; they usually are irreplaceable, as one can never take an individual back to the point in time when the specimen was originally collected. The more a collection is used, the greater its value, because it is possible to examine different causal pathways without collecting and testing for the already determined parameters. However, even if the collection is of a renewable resource, for example, bacteria, there is a cost associated with sharing with

---

**TABLE 12.4** Ethical Issues Associated with Biorepositories

- Responsible custodianship
- Informed consent
- Privacy protection
- Access to biospecimens and data
- Intellectual property and resource sharing

others, not just the monetary costs of packing and shipping which can be charged to the user, but costs associated with loss of intellectual property, and concerns that data will be misinterpreted or misused or used without attribution. A thoughtfully collected and well-maintained collection, annotated with medical history and sociodemographic information, can result in inventions or data that have commercial value. Therefore it is not surprising that investigators are reluctant to share biological specimens. However, if a collection is not used it has no value, and the costs associated with maintenance are for naught.

The principle of responsible custodianship requires proper handling and storage of biospecimens. This entails having appropriate personnel, standard operating procedures, and appropriate quality control and quality assurance measures. While this may be fairly straightforward to provide during the course of a study, it may be more difficult following the end of grant funding. If the investigator intends to store specimens for future study, plans should be made to ensure that biorepository specimens will be maintained and made available for future scientific use, while protecting the confidentiality of associated data.

## 12.5.2 Informed Consent

Many biorepositories are established in conjunction with specific studies and then subsequently maintained; others are established specifically as repositories for future research. Thus the informed consent obtained may be different, which can pose ethical issues regarding subsequent use. If the tissues are linked to personal information, such as medical history, the data fall under privacy regulations (HIPAA), which has additional requirements. For example, HIPAA requires obtaining permission for each use; although at this writing how HIPAA is implemented for biomedical research is under review. Investigators might ask participants if they can be recontacted for permission to use the specimen in future studies. This can be problematic, however, because participants may be lost to follow-up or resent recontacting. If the data are made anonymous, that is, the linkage between identifying information and the specimen is destroyed, using the specimens for other purposes may qualify for an exemption.

Specimen use should be consistent with the informed consent, and there should be a process in place for participants to withdraw consent and request that a donated specimen be destroyed. This can be challenging if the specimen was distributed to several investigators or if the linkage between identifying information and the specimen was destroyed. If the specimen has already been distributed to others, or subjected to testing, must these data be destroyed? According to the National Cancer Institute (NCI) best practices, these data need not be withdrawn, but all future testing is prohibited. Further, the specimen need not be retrieved from other investigators, but the other investigators should be notified.

When tissues are collected as part of repositories for future research, it is virtually impossible for participants to be informed of the future use of their specimens. Large-scale biobanks are intended to follow individuals prospectively throughout their lives and to be a societal resource, meaning the future uses are unpredictable.[19] Examples of biobanks include the Iceland Biobank, administered by the company DeCode, and the United Kingdom Biobank. The United States is engaged in setting up a large scale biobank following children and their mothers over time, called the National Children's Health Study. The risks and benefits of such large-scale public investments cannot be evaluated in terms of respect for individual autonomy, but in terms of larger benefits to society relative to individual risks. The balance of societal benefits to individual risks is a central ethical concern of public health.

## 12.5.3 Privacy Protection

Specimens are more valuable if there is associated information, such as the sociodemographics of the participant, medical history, and clinical information. However, this information also makes it more difficult to protect an individual's privacy. In addition, there are issues regarding to whom current and future results should be released.

191

Information might jeopardize the individual's ability to get health insurance, for example, because an insurance company might use the information to demonstrate a preexisting condition. Biorepositories should have clearly articulated policies about protection of privacy of participants and confidentiality of the data.

It is possible that continued study of biorepository specimens will lead the investigator to identify individuals at high risk of disease or with an early disease stage. If it is possible to identify the individual, there is an implied responsibility to the person who donated the specimen. However, the responsibility of the investigator depends upon the timing of the discovery, the particular disease, and what was stipulated within the informed consent document. The investigator has a responsibility to do no harm; if early detection makes no difference in disease outcome, informing the participant may cause harm. If detection occurred many years after the disease was likely to have occurred, informing the participant is similarly unhelpful. However, it may be stipulated in the informed consent document that the participant will be informed of study results. This clause is typically included in informed consent documents, because it provides some incentive for participation. In this case, the investigator is obligated to find an appropriate method of imparting the information. One mechanism is to notify the participant's treating physician. However, if the participant has no treating physician, this can be problematic. Obtaining guidance from one's institutional review board in these situations can be extremely helpful; it also provides institutional support for whatever is the ultimate decision.

## 12.5.4 Intellectual Property and Resource Sharing

In the United States, most public and private funding agencies award grants to institutions, not individuals. Any equipment purchased or material collected is thus the property of the institution, not the individual investigator. If an investigator changes institutions, the first institution must agree to the transfer of materials and funds. If the investigator retires or dies, the materials remain the property of the institution. Unless the appropriate infrastructure has been put in place, with associated documentation, future use of the material may be compromised or lost. Biological materials can be costly to store, and associated administrative structures are not without cost, so that without an advocate or formal contracts to the contrary, the pressures will be to dispose of materials rather than shoulder the burden of maintenance and making the material accessible.

There have been a number of lawsuits regarding the use of patient specimens to develop a commercial product.[18] One issue is whether the specimen donor has a right to any financial benefits that may have been obtained from a product developed from the specimen. Although the court ruled no, it would be prudent for an investigator to clarify in an informed consent document that this is the case. A second issue is whether the donor has any control over how the specimen is used. This has been a particular problem with genetic data; once genetic sequence has been determined, it can be analyzed for relationships with any characteristics available to the investigator, which may not be acceptable to the respondent. Genetic testing can identify relationships, such as that a child's father was not the biological father, or that a family origin myth is correct or incorrect. Ensuring privacy of information and including in the consent documents clarification of to whom results can be released is thus essential. Finally, as technology advances, participants might wish return of the specimen for their personal use. For example, it is already possible to grow tissue from cells; a participant suffering from an injury or illness might wish to avail themselves of this technology. As these technologies move out of the realm of science fiction, clarifying whether specimens will be returned, and to whom, should be included within the consent document.

If a collected specimen is associated with a specific study, to whom it will be distributed for testing, and how it will be tested should be detailed in the study protocol and standard operating procedures. The protocol may specify storage of replicates in case retesting is needed. Following the end of the study, the remaining specimens might be disposed of, but often are saved because of their potential for future study. Ideally, procedures will be set in place to make others aware of the resource, and to share it with others in a way consistent with the original informed consent of participants. This may entail setting up a review committee to vet requests.

## 12.6 PUBLIC DATA ACCESS

There is power in numbers; pooling data from different studies increases the power to detect associations. Making data freely available enables explorations that would be difficult for an investigator to conduct alone, such as comparing genomes across organisms, speeding scientific discovery. While freely available, there is a monetary cost to making data public. There is also the risk of compromising the privacy of individuals whose data are included. Informed consent documents routinely promise that any data made freely available will be made anonymous. As databases grow, so does the ability to link them together, increasing concerns that even anonymous data can be identified. Genetic sequence data is a particular concern. Further, if the data can be linked to identifying information, it places the investigator in a potentially awkward legal position if the individual in question is under criminal investigation. Thus, although there are substantial benefits to making data generated freely available, there are some potential drawbacks.

It is now the policy of the National Institutes of Health (NIH) in the United States that data from any study whose annual direct costs exceed $500,000 per year must have a data sharing plan. To date, data sharing has worked most effectively with sequencing data from the Human Genome Project, as any investigator can deposit sequence data in publicly accessible databases maintained by the NIH. Similar databases are available for sequence data from other organisms and for other types of omics data. With these developments, standardized procedures for documentation have been developed, and most journals require inclusion of the accession number, which is obtained when sequence is submitted to a public database, in published material. For more information on omics databases, see the National Center for Biotechnology Information (NCBI) website at http://www.ncbi.nlm.nih.gov.

For other types of public access, the burden of making the material available is borne by the investigator and associated institution. It is the policy of some scientific journals that authors must share materials. For example, the American Society of Microbiology journal, Antimicrobial Agents and Chemotherapy, requires "the authors agree that, subject to requirements or limitations imposed by laws or governmental regulations of the United States, any DNAs, viruses, microbial strains, mutant animal strains, cell lines, antibodies, and similar materials newly described in the article are available from a national collection or will be made available in a timely fashion, at reasonable cost, and in limited quantities to members of the scientific community for noncommercial purposes. The authors guarantee that they have the authority to comply with this policy either directly or by means of material transfer agreements through the owner." This can put a considerable burden on an individual investigator, as the costs of technician time, packing, and shipping can be considerable – even if the requester assists with the costs.

Until recently, data sharing was not a standard feature of epidemiologic studies. Although there are many large-scale studies that have been worked on by many individuals, making data publicly available has not been the norm. Barriers to data sharing include how to implement mechanisms to protect the intellectual investment of participating investigators, the costs of making data publicly available in terms of documentation and administration and who bears these costs, concerns regarding protecting privacy of study participants, and potential decreased participation for studies with an open-ended purpose. Further, epidemiologic studies are like custom designed machines. Every data set has its peccadillos that may not be easily discerned from even the best documentation. Having an individual familiar with the data set is essential for appropriate interpretation. Data

| TABLE 12.5 | Principles of Sharing Epidemiologic Data. Policy Statement of the American College of Epidemiology |
|---|---|

- Science and the utility of the epidemiologic approach to generating new knowledge benefit when data are shared with other investigators.
- Whenever possible, secondary users of existing data sets should work with the original investigators. It is desirable that those who collected or produced the original data be co-authors or be given the right of refusal.
- The rights and privacy of people who participate in epidemiologic research must be protected and the confidentiality of research data should be safeguarded at all times.
- Informed consent for future use of data beyond the original study should be obtained at the time of the original study.
- Data should be made available for sharing as soon as possible after publication of the main results of a study.
- Archiving of data sets, especially those that are unique, is to be encouraged when possible and appropriate.
- The costs of data sharing may be borne by the entities funding the original research or those supporting the secondary use of the data.
- Data sharing arrangements should take the time requirements, beyond pure costs, into account.
- Data sharing may not be appropriate in some circumstances.

Source: Adapted, with permission, from American College of Epidemiology[20] (2002).

must be de-identified before making it available; if posted on a website, the website must be maintained. Alternatively, if made available by request, the data must be readable by current computer formats. Few computers are available that can read data stored on tapes or floppy disks, the standard formats in the fairly recent past. Finally, data sharing must respect any third-party agreements in place. To address these concerns, the American College of Epidemiology developed a policy on data sharing.[20] The key points of this policy are shown in Table 12.5. Note that the last principle is that in some circumstances, data sharing may not be appropriate. This may be the case for past studies not designed with an eye toward making data public in the future. However, it is difficult to argue that data from a properly conducted study should not be shared. Data sharing enables fuller use of existing data and provides a strong incentive to adhere to the highest standards of data conduct, analysis, and documentation. Adhering to these standards improves all science.

## 12.7 SHIPPING OF MATERIALS BETWEEN INSTITUTIONS

In the past, it was not uncommon for researchers to share biological materials informally, without oversight, although there were regulations in place with regard to packing and labeling for shipping. The events of September 11, 2001, and subsequent bioterrorist attacks led to the development and implementation of rules and regulations regarding sharing of biological materials. U.S. regulations regarding packing and transport requirements can be found on the Centers for Disease Control and Prevention (CDC) website (http://www.cdc.gov). Permits are required for importing infectious agents; these can be obtained from the CDC. Transfer of research materials, such as infectious agents, between institutions requires the signing of a Material Transfer Agreement, also available from the CDC. Readers are cautioned to check with the appropriate committees at their institution before shipping or requesting biological materials from other investigators, in order to ensure proper procedures are followed.

## 12.8 PROTECTION OF ANIMAL SUBJECTS

There are excellent reasons for conducting research on animals. Animals can be kept in controlled environments, which minimizes the effects of environmental variables on results; there are animals that are bred to be at high risk of a disease of interest, minimizing the sample sizes required; and there are genetic lineages of animals so that effects of genetics can be minimized. Because of the large number of uncontrolled (and uncontrollable) variables in human studies – even in clinical trials – corresponding human studies would have be substantially larger and would be of significantly greater cost. However, there is an ethical cost in using animals; most significantly, animals are usually euthanized at the end of the experiment. Moreover, animals cannot give their consent. Therefore animal studies should not be undertaken lightly, and investigators planning studies in animals are subject to review and approval. Only studies that clearly demonstrate that the studies are necessary to forward knowledge and that the animals will be properly cared are approved.

Similar to human subject research, there have been past abuses that led to current regulations. For example, Pasteur believed that animals could not feel pain, and conducted surgeries and procedures without the benefit of anesthesia or other safeguards. To prevent researchers from inflicting unnecessary pain and suffering in the name of science, animal research is heavily regulated. Regulations specify what constitutes humane handling, care, treatment, transportation, veterinary care, and record keeping requirements.

Each institution must have a committee to review all care and use of animals in research, usually called the Institutional Animal Care and Use Committee (IACUC). The IACUC must include at least a laboratory animal veterinarian, a researcher who uses laboratory animals, a nonscientist, and a member of the public not affiliated with the institution. The IACUC reviews proposed research projects to ensure that the project (1) has the potential to allow us to learn new information, (2) will teach skills or concepts that cannot be obtained using an alternative, (3) will generate knowledge that is scientifically and socially important, and (4) is designed such that animals are treated humanely.[21]

All investigators engaging in animal research must receive appropriate training in protection of animal subjects. All study protocols must be submitted to the appropriate review committee. Facilities housing animals are subject to regulation and oversight. Because of the considerable administrative burdens, most institutions have special animal facilities. Investigators planning on conducting animal studies are directed to contact their local facility and review committee for determining the appropriate procedures.

## 12.9 BIOLOGICAL SAFETY

Biological safety means using infectious and other hazardous materials safely in the laboratory, and reducing or eliminating the exposure of laboratory personnel, other personnel, and the outside environment. The primary principle is that of containment. Biological materials are contained by use of good technique, safety equipment, and vaccination of personnel. The degree of containment required depends on the biological material; these are classified into four levels with four being the highest level (Table 12.6). Handling of diagnostic specimens and recombinant DNA generally is carried out in laboratories with equipment consistent with levels 1 or 2. Infectious agents that cause serious disease and are transmitted by aerosols require safety levels 3 or 4. It is the responsibility of the investigator to ensure that the appropriate safety levels are met.[22]

Modern molecular tools often use recombinant DNA. Recombinant DNA includes molecules constructed outside of living cells by joining natural or synthetic DNA segments to DNA molecules that can replicate in a living cell, or molecules that result from their replication. Also included are host vector systems, plasmids, and viruses. Thus cloning, transformation, and other routine procedures in molecular biology fall under this rubric. Institutions that receive NIH funding must have an Institutional Biosafety Committee that ensures that all recombinant DNA research conducted at the institution meets NIH guidelines, regardless of whether the research is funded by the NIH. The Institutional Biosafety Committee is similar to the Institutional Review Board in that it has both oversight and reporting duties. It also has a specified composition that must include two outside members as well as those from the institution with appropriate expertise. The purview of the Institutional Biosafety Committee includes recombinant DNA, select agents, and human gene transfer studies.

**TABLE 12.6 Biosafety Levels (BSL) and Practices for Containment**

| BSL | Procedure | Practices to Follow |
|---|---|---|
| 1 | Not known to consistently cause disease in healthy adults | Standard microbiological practices |
| 2 | Associated with human disease, hazard = percutaneous injury, ingestion, mucous membrane exposure | BSL-1 practices plus:<br>• Limited access<br>• Biohazard warning signs<br>• "Sharps" precautions<br>• Biosafety manual defining any needed waste decontamination or medical surveillance policies |
| 3 | Indigenous or exotic agents with potential for aerosol transmission; disease may have serious or lethal consequences | BSL-2 practices plus:<br>• Controlled access<br>• Decontamination of all waste<br>• Decontamination of lab clothing before laundering<br>• Baseline serum |
| 4 | Dangerous/exotic agents that pose high risk of life-threatening disease, aerosol-transmitted lab infections; or related agents with unknown risk of transmission | BSL-3 practices plus:<br>• Clothing change before entering<br>• Shower on exit<br>• All material decontaminated on exit from facility |

Source: Adapted, with permission, from U.S. Department of Health and Human Services[22] (2007).

Select agents are microorganisms or specific toxins that are identified as select by the U.S. government (the list of select agents for the United States is available on the website of the National Select Agent Agency, at http://www.selectagents.gov/Select%20Agents%20and%20Toxins%20List.html). Handling of select agents is subject to rules and regulations intended to protect the health and welfare of those working with the materials and the wider community that might be exposed if the materials are not handled properly. Molecular techniques also entail working with noninfectious but hazardous materials. To protect workers and ensure that government regulations regarding workers (Occupational Safety and Health Administration [OSHA] regulations), each institution has an OSHA office. The researcher should check with the Institutional Biosafety Committee (IBC) and OSHA office in his or her own institution to make sure that his or her laboratory is compliant.

## 12.10 PROFESSIONAL ETHICS

Unlike the topics touched on earlier, professional ethics for epidemiology and public health are not regulated. However, there is a code of ethics for the practice of epidemiology that has been adopted by several epidemiology societies[23] and for public health professionals, which has been approved by the American Public Health Association.[24] These general principles are summarized here, with particular examples relating to molecular epidemiologic studies.

### 12.10.1 Do No Harm

The relationship of an epidemiologist to a community can be likened to that of a physician to a patient: there is an obligation to do no harm, to provide benefit, to respect the rights and welfare of study participants and community members, to disclose conflicts of interest, and to honor all obligations to the community[25] (Table 12.7). Some of these principles overlap with the principles discussed in Section 12.3. The obligation to do no harm entails minimizing risks relative to expected benefits, and limiting research projects to ones whose results are likely to provide benefit, either in furthering scientific knowledge with potential short- or long-term benefit, or providing information for use in policy-making or planning public health interventions.

### 12.10.2 Provide Benefit

Benefits of epidemiologic studies are generally to society, rather than directly to the individual, but studies that do not have the potential to provide benefit to society should

---

**TABLE 12.7** Ethical Code for Molecular Epidemiologic Studies

- Do no harm
- Provide benefit
- Respect the rights and welfare of study participants and community members
- Disclose conflicts of interest
- Honor obligations to the community

---

not be conducted. Respecting the rights and welfare of study participants and community members requires protecting privacy and confidentiality, and obtaining informed consent. Disclosure of conflicts is intended to alert the user of the information to (presumably) unconscious biases that might have occurred during study design, conduct, analysis, and presentation. These conflicts are not only financial, but currently only financial conflicts must be disclosed. Honoring all obligations to the community means engaging the community during the design and conduct of the study; informing them of study results and of any risks to their health and welfare that might have been detected during study design, conduct, or analysis; and respecting their contributions to the research effort.

### 12.10.3 Respect the Rights and Welfare of Study Participants and Community Members

Individual and institutional study participants should be treated with respect, keeping in mind that they are doing the investigator a favor. The investigator should minimize any personal discomfort or inconvenience associated with participation; study conduct should strive to enhance rather than disrupt normal community functioning. For example, when working with a patient clinic, the investigator should identify a time in the typical patient trajectory through the clinic when study recruitment will not disrupt overall patient flow. This might be when the patient is waiting for care. The study might enhance the patient experience by providing parking services for study participants or bus passes, or facilitate clinic functioning by having study personnel assist with normal clinic tasks that might also enhance the study. Study personnel might triage urgent care visits while simultaneously identifying potential study participants.

### 12.10.4 Disclose Conflicts of Interest

Conflicts of interest occur "whenever a personal interest or role obligation conflicts with an obligation to uphold another party's interest, thereby compromising normal expectations of reasonable objectivity and impartiality."[26] A financial interest or strongly held belief can influence publication of study results; conflicts can and do adversely affect scientific judgment, either consciously or unconsciously.[25] The 2009 Institute of Medicine report addressed conflicts of interest for clinicians,[27] and similar conflicts occur within epidemiology. Disclosures of funding sources provide readers of published papers a method of discerning if a conflict might have occurred, but do not effect an author's decision to publish a paper or how results are presented or an argument framed.[28] Funding source is not the only source of conflict; someone working on behalf of industry may be less biased than someone working on behalf of a public interest group. Ethically, we should judge the science presented on its own merits, critique errors in design, conduct and analysis, and avoid ad hominem attacks based on the author's place of employment or source of funding. Nonetheless, gifts and financial support do generate a feeling of obligation to those providing the gifts or support, which may cause the recipient to subconsciously soften or emphasize certain findings in published reports or lose interest in certain lines of research. Similarly, an individual who has built his or her scientific reputation on a certain hypothesis may be more critical of research that does not support his or her beliefs. We should strive to identify the kernel(s) of truth that might be found in all studies conducted by well-meaning individuals, as we strive to increase our understanding of important public health problems.

### 12.10.5 Honor Obligations to the Community

Individuals studying populations have an obligation to the study population[26] during all aspects of a study, including the design, conduct, analysis, and reporting of study results. When possible, the design phase should include engaging stakeholders. Understanding the concerns of the study population and contextual factors will ultimately improve study design and conduct. Community members are sources of important information about how to best contact and engage other community members, which increases participation. They can also give guidance on contextual factors that might mediate findings that should be measured, such as social or behavioral variables, assist in generating alternative hypotheses, and give insight into study results. An excellent example of this is recounted by Elisabeth Pisani,[29] when she described conducting a seroepidemiology study among commercial sex workers in Thailand. Recruiting sex workers available during peak business times ended up with a very biased sample and an underestimate of HIV prevalence.

## References

1. N.E. Kass, An ethics framework for public health, Am. J. Public Health 91 (11) (2001) 1776–1782.

2. U.S. Department of Health & Human Services, Office of Extramural Research, Glossary & Acronym List [website], Available at: http://grants.nih.gov/grants/glossary.htm (Updated 19.2.10; accessed 17.8.10).

3. A.L. Fairchild, R. Bayer, Public health. Ethics and the conduct of public health surveillance, Science 303 (5658) (2004) 631–632.

4. N. Riccardi, T. Monmaney, King/Drew medical research suspended. Voluntary action comes after regulators find two dozen violations of rules designed to protect human subjects and ensure credibility at south L.A. facility, Los Angeles Times (April 27, 2000).

5. R.L. Penslar, U.S. Department of Health & Human Services: Office for Human Research Protections (OHRP): IRB Guidebook [Internet], U.S. Department of Health and Human Services, Available at: http://www.hhs.gov/ohrp/irb/irb_guidebook.htm (accessed 17.8.10).

6. H. Jonas, Freedom of scientific inquiry and the public interest. The accountability of science as an agent of social action, Hastings Cent. Rep. 6 (4) (1976) 15–18.

7. A.L. Caplan, Twenty years after. The legacy of the Tuskegee Syphilis Study. When evil intrudes, Hastings Cent. Rep. 22 (6) (1992) 294.

8. H. Edgar, Twenty years after. The legacy of the Tuskegee Syphilis Study. Outside the community, Hastings Cent. Rep. 22 (6) (1992) 324.

9. P.A. King, Twenty years after: The legacy of the Tuskegee Syphilis Study. The dangers of difference, Hastings Cent. Rep. 22 (6) (1992) 354.

10. H.K. Beecher, Ethics and clinical research, N. Engl. J. Med. 274 (24) (1966) 1354–1360.

11. S. Spotswood, Department of Veterans Affairs Office of Inspector General. Healthcare Inspection: Human Subjects Protections Violations at the Central Arkansas Veterans Healthcare System Little Rock, Arkansas. VA Office of Inspector General. Washington, DC. August 6, 2008. Available at: http://www4.va.gov/oig/54/reports/VAOIG-07-03042-182.pdf

12. T.L. Beauchamp, J.F. Childress, Principles of Biomedical Ethics, sixth ed., Oxford University Press, New York, 2009.

13. R.K. Zimmerman, HPV vaccine and its recommendations, J. Fam. Pract. 2007 56 (Suppl. Vaccines 2) (2007) S1–5 C1.

14. National Institutes of Health, Office of Human Subjects Research, Regulations and Ethical Guidelines [website], Available at: http://ohsr.od.nih.gov/guidelines/index.html (Updated 3.9.09; accessed 17.8.10).

15. R.B. Ness, Influence of the HIPAA Privacy Rule on health research, JAMA 298 (18) (2007) 2164–2170.

16. R. Hakimian, S. Taube, M. Bledsoe, et al., 50-State Survey of Laws Regulating the Collection, Storage, and Use of Human Tissue Specimens and Associated Data for Research. National Cancer Institute Cancer Diagnosis Program, National Cancer Institute Available at: http://cdp.cancer.gov/humanSpecimens/survey/index.htm

17. National Cancer Institute, National Institute of Health, U.S. Department of Health and Human Services: National Cancer Institute Best Practices for Biospecimen Resources [pdf], Available at: http://biospecimens.cancer.gov/practices, 2007 (Updated 12.4.10; accessed 17.8.10).

18. J.B. Vaught, N. Lockhart, K.S. Thiel, et al., Ethical, legal, and policy issues: Dominating the biospecimen discussion, Cancer Epidemiol. Biomarkers Prev. 16 (12) (2007) 2521–2523.

19. G. Williams, Bioethics and large-scale biobanking: individualistic ethics and collective projects, Genomics Soc. Policy 1 (2) (2005) 50–66.

20. American College of Epidemiology, Principles of Data Sharing. Policy Statement on Sharing Data from Epidemiologic Studies, American College of Epidemiology, Raleigh (2002) 2–5.

21. American Association for Laboratory Animal Science, Institutional Animal Care and Use Committee: About Us [website], Available at: http://www.iacuc.org/aboutus.htm (Updated 22.1.09; accessed 17.8.10).

22. U.S. Department of Health and Human Services, Public Health Service, Centers for Disease Control and Prevention, et al., Biosafety in Microbiological and Biomedical Laboratories, fifth ed., Government Printing Office, Washington, DC: US, 2007.

23. American College of Epidemiology, Ethics guidelines, Ann. Epidemiol. 10 (8) (2000) 487–497.

24. Public Health Leadership Society. Principles of the ethical practice of public health [website], National Network of Heath Institutes, Available at: http://phls.org/CMSuploads/Principles-of-the-Ethical-Practice-of-PH-Version-2.2-68496.pdf (Updated 11.3.10; accessed 11.3.10).

25. S.S. Coughlin, Ethical issues in epidemiologic research and public health practice, Emerg. Themes Epidemiol. 3 (2006) 16.

26. C.L. Soskolne, A. Light, Towards ethics guidelines for environmental epidemiologists, Sci. Total Environ. 184 (1-2) (1996) 137–147.

27. Committee on Conflict of Interest in Medical Research Education and Practice, Institute of Medicine, in: B. Lo, M.J. Field, (Eds.), Conflict of Interest in Medical Research, Education, and Practice, National Academies Press, Washington, DC, 2009.

28. N. Pearce, Corporate influences on epidemiology, Int. J. Epidemiol. 37 (1) (2008) 46–53.

29. E. Pisani, Wisdom of Whores, 1st American ed., W.W. Norton & Company, Inc., New York, 2008.

# CHAPTER 13

# Future Directions

This volume includes many examples of the applications of molecular techniques in infectious disease epidemiology and public health that were current in 2010. In this final chapter, the focus is on potential future applications. At this writing, many of these ideas are already under development, so the reader is directed to the scientific literature for further information. Applications under most rapid development stem from the ability to cheaply and rapidly sequence the genome of humans and microbes, and to sequence genes in entire microbial communities. This capability has a high potential to transform the future of epidemiology on a variety of levels. Readily available genetic data can enhance surveillance, the tracing of transmission flows, and predictions of future disease spread. Further, combining genomics or proteomics with microarray and microfluidic technology enables rapid identification of viruses, bacteria, and fungi. Examination of human–microbe and microbe–microbe interactions is increasingly feasible using genomic and proteomic data. Regardless of the application there will be ample work when the results are applied to populations. New measures require new decision rules. New measures must be mapped to the existing standards, and an assessment made of whether the new measure evaluates the same, an alternative, or overlapping construct of the standard, and how a new measure might influence decisions regarding the need for investigation and intervention. Analysis and interpretation of the measure may also require methodological development.

## 13.1 METHODOLOGICAL DEVELOPMENT

New tools need to be validated, the reliability determined, and the results interpreted. The steps for determining validity and reliability are the same regardless of the measure and were described in detail in Chapter 8. Interpretation is more complex and may require new methods to satisfactorily translate the outcome of a new technology to a meaningful result. This is motivating microbiologists to learn statistics and epidemiology and fueling the field of bioinformatics. Epidemiologists are used to dealing with large data sets, but we will need new strategies to effectively use the vast amounts of data generated from genomic, metagenomic, and proteomic studies. My view is that this will also require a paradigm shift. Infectious disease epidemiology will need to move from the paradigm of empiricism to a more theoretical framework in order to fully explore the complex interactions that result in the transmission, pathogenesis, and evolution of infectious agents. A shift to a more theoretical framework is already occurring, particularly among research focused on transmission systems (see for example Li and associates[1] [2009]).

Traditional epidemiology, embodied by outbreak investigation, is very empirical. To someone trained in evolutionary biology or ecology where experiments are designed to test specific hypotheses that arise from the underlying theory and models, the contrast is stark. Epidemiology revolves around determining the cause, and usually the most proximate cause, of observed phenomena. Determining a cause enables the epidemiologist to prevent disease by addressing that cause, and in so doing, protect the public's health. Because the studies

Molecular Tools and Infectious Disease Epidemiology.
© 2012 Elsevier Inc. All rights reserved.

are primarily observational, there can be significant potential for error in identification, and much of epidemiologic training consists of learning methodologies aimed at reducing those errors. Nonetheless, very successful public health interventions can and have been based on a very limited understanding of the mechanisms behind disease transmission and pathogenesis. Because human health is at stake, it is generally deemed better to act upon limited information rather than to wait until certainty is achieved. A classic example is the (probably apocryphal) removal of the handle on the Broad Street pump by John Snow.

Public health interventions to ensure clean water and air, protect our food supply, and limit crowding have contributed substantially more to our overall health than any medical interventions, but were implemented based on quite simple observations. The pragmatic approach does have limitations, because there are unintended consequences of every intervention. In comparison to the scourges of the past, the current scourges are more subtle and the correct path to prevention is often less clear. To accurately predict the effect of a proposed intervention requires a more complete understanding of the mechanisms behind the phenomenon of interest and the interactions occurring in different components of the overall system. This requires building a theoretical framework. Fortunately, there are already well-developed or developing theoretical frameworks from which epidemiologists can draw.

One theoretical foundation can be found in ecology, which focuses on interactions and competition among and within populations and communities. The focus on interaction provides an alternative to reducing an infectious disease process to the actions of a single agent with its host and environment. Ecology is essential for understanding the epidemiology of co-infections, microbiota, and for making sense of the genetic diversity observed among strains of a single species. The benefits of integrating ecology with epidemiology have already been demonstrated by a subdiscipline of ecology, disease ecology. Another foundation is evolutionary theory. Infectious disease epidemiologists often refer to evolution as an explanation for observed phenomena, such as selection for antibiotic resistance, but rarely use evolutionary theory as a framework for developing research questions. Evolutionary theory is arguably the best developed scientific theory and provides alternative ways to frame questions as well as explain phenomena. Considering how human behavior selects for or against transmission or virulence factors provides a robust framework for enhanced prediction of the unintended consequences of a public health intervention. Lastly, a systems approach provides a way to understand nonlinear, dynamic, adaptive systems, which describes most biological processes. Systems approaches use mathematical models to predict the course of an outbreak, evaluate different policies, and explore mechanisms that might explain observed phenomena.[2] The enhanced measurement enabled by molecular tools makes it easier to test evolutionary and ecological theories in vivo, and estimate essential parameters required for prediction models. The resulting synergies and increased understanding that emerges from a merger of practice and theory opens an exciting range of possibilities for the molecular epidemiologist that should ultimately lead to enhanced ways of protecting the public's health.

## 13.2 SURVEILLANCE

Infectious disease surveillance has already adapted some molecular methods to great effect. Pulsed-field gel electrophoresis transformed surveillance for food-borne illness (see Chapter 6), and spoligotyping transformed tuberculosis surveillance. Even more sensitive and specific techniques, based on genotyping, are becoming more affordable, which, with appropriate infrastructure support, will further enhance surveillance. Genotyping methods, such as multilocus sequence typing (MLST), have great advantages over gel-based methods. The biggest advantage is that the results are portable, they can be summarized with a number, and they are less subject to variation when conducted in different laboratories or by different operators. Gel-based typing methods vary from machine to machine and operator to operator, requiring repeated testing of potentially similar strains, which slows the identification of outbreaks and elimination of pseudo-outbreaks.

Genetic sequences are generally more discriminatory than gel-based techniques. Sequence data also facilitate phylogenetic analyses, which can be very helpful in sorting out the origin of pathogens. One application is determining the source of contamination of recreational water. Fecal coliform can live for long periods in sand; contact with sand next to recreational water is associated with an increased risk of gastrointestinal disease.[3] However, there are many possible sources of fecal coliform, from human to water fowl, to upstream contamination, such as runoff from feed lots. Potentially pathogenic viruses can also live in water. Distinguishing between animal and human contamination, and the source of animal contamination (water fowl or runoff from feed lot), leads to more effective interventions, enhanced public relations, and increases the probability that future outbreaks will be prevented.

Disease transmission is generally local. Although the potential for global spread with modern transportation is high, and there is increasing evidence that microbes can move vast distances in dust storms, most disease is transmitted very locally. Surveillance can take advantage of this phenomenon; more sensitive and specific typing enables identification of geographic areas with high levels of ongoing transmission of disease and of the networks of disease spread. Identification of areas with high levels of ongoing transmission is critical for targeting interventions. Though this idea is not new, what is new is the ability to separate different ongoing chains of transmission. Molecular typing has revealed that even in geographic areas with high endemic rates, there may be separate transmission systems associated with different sociodemographic groups. This was demonstrated by Haase and associates[4] (2007) using IS6110-RFLP and spoligotyping (described in Chapter 6) for tuberculosis in Canada. Tuberculosis rates were high among immigrants, and citizens living in the same area had higher rates of tuberculosis than citizens living elsewhere, although lower rates than that found among immigrants. The assumption was that immigrants were the source of high tuberculosis rates among citizens. This turned out not to be the case. Tuberculosis among immigrants was essentially independent of transmission among nonimmigrants.[4] Similarly, the high rates of gonorrhea among men who have sex with men were presumed to be a source of disease among heterosexuals, with bisexual males acting as a bridge. However, studies of gonorrhea transmission in the Netherlands found that the strains circulating among gay men were different from those circulating among heterosexual couples with little or no overlap.[5] One can envision this application of molecular typing expanded to the control of a variety of diseases with endemic circulation. A Texas health department applied these ideas to tuberculosis control, greatly increasing yields and decreasing ongoing transmission.[6] These efforts require not only the ability to type strains, but also to analyze the data, link with geographic area in a timely manner, and develop the political will to intervene.

Rather than limiting surveillance to monitoring of known diseases, surveillance can be devised to best detect sources of future diseases. Diseases such as AIDS, which appeared in the United States sometime in the late 1970s, West Nile virus, which appeared in the United States in 1999, monkey pox virus, which has been in the United States since 2003, and antibiotic resistance in a variety of human pathogens have all been rapidly disseminated. Of particular concern is the potential for diseases to jump from wild animals to humans and cause serious disease. Surveillance in areas where there is frequent interaction between humans and wild animals have been targeted for monitoring.[7] Surveillance might include monitoring humans who have close interactions with animals that have the highest potential for acquiring a zoonotic disease that might jump to humans. Systems where there is pressure for evolution of infectious agents, such as waste water treatment plants, might also require monitoring. These types of surveillance systems are already in place in some areas.

Worldwide monitoring is essential, as the global economy makes it possible for the rapid dissemination of disease. In addition to more exotic diseases, nonculture techniques coupled with genotyping makes it possible to more rapidly identify new strains of well-known scourges like influenza, implement interventions, and monitor disease spread. Timeliness can be very important, as illustrated by the 2009 influenza A H1N1, which was first identified in Mexico in February 2009. By October 2009 there were cases reported worldwide and in all 50 states of the United States.

The richer data that result from surveillance using modern molecular methods enhances our ability to accurately predict the course of an outbreak or evaluate different policies using models. Collecting, analyzing, and disseminating the data is not without costs. In addition to adding new laboratory equipment and more discriminating assays, personnel may need to be retrained, surveillance reports updated, and users of surveillance data educated in the interpretation of the new measure. Education should include an introduction to the theory behind the new molecular method and its interpretation, and how it changes current operating procedures.

## 13.3 TRANSMISSION

Infectious diseases are transmitted between humans directly, through person-to-person contact, and indirectly, via the environment (air, food, water, or surfaces) or a vector (such as a mosquito). Many bacteria are found on the skin, and as skin cells are sloughed into the environment the bacteria are also shed. Similarly, bacteria and virus in the nose are aerosolized and deposited in the environment following sneezing or coughing. Depending on the microbe, the dose, environmental parameters, and the characteristics of what is contacted next, touching contaminated surfaces can transmit the microbes; airflow might also lead to transmission of infection. Although many microbes can be communicated via multiple modes, during an outbreak, one transmission mode usually predominates. Estimates of the relative contribution of different modes on ongoing transmission of a particular microbe are rare. Molecular typing makes it feasible to track the origin of specific microbial strains and estimate the contribution of different modes to overall transmission rates. Of particular interest is estimating these parameters for transmission within hospitals and long-term care facilities.

Individuals who are already ill or in a hospital tend to be more susceptible to infection. They have wounds from surgical procedures, trauma, and insertion of medical devices; all of these provide a portal of entry for infection. Wound- and device-associated infections are extremely common. It is assumed that the source of these infections is from self-inoculation or inoculation by hospital personnel, although some infections may be acquired from the environment. Determining the relative contribution of different sources is very important for minimizing hospital-associated infections. Current focus is on the role of hospital workers as transmitters of infection, but if self-inoculation is a major contributor then interventions might be developed to limit opportunities for it to occur. Among hospital patients, the presence of *Staphylococcus aureus* in the anterior nares (the nose) greatly increases risk of wound infection. One prevention strategy is to screen for *S. aureus* upon admission, followed by treatment of those who are positive. Other strategies might be desirable, because screening and treatment will undoubtedly increase pressure for antibiotic resistance in *S. aureus* in an environment where selection for resistance is already high. One possibility would be to explore factors that reduce *S. aureus* dispersal; an intervention might enhance *S. aureus* biofilm formation and stability of colonization in the nose.

Health care interventions frequently involve cleaning. Whether the cleaning removes all microbes is unclear, but it reduces microbial presence. Though cleaning definitely reduces risk of infection from existing microbes in a wound, if it disrupts normal microbiota, it can enhance the potential for invasion by new microbes. The microbes might come from outside the cleaned area, from health care personnel, the patient (e.g., while manipulating the wound), or from the environment. If the process were better understood, one could imagine a day where there is a cleaning followed by an inoculation step; beneficial microbes might be introduced to increase the chances that healing will be successful.

Horizontal gene exchange is an important mechanism used by bacteria to acquire virulence factors. However, it is often not the host bacteria but the exchanged genetic material (known as mobile genetic elements) that decides when and how the material is exchanged. Tracing the transmission of mobile genetic elements and identifying the determinants of exchanges is epidemiology on the molecular level. Studies identifying the determinants of gene exchange can provide important information for guiding development of new treatment and intervention strategies. For some mobile genetic elements, exchange tends to occur during periods of stress, such as exposure to antibiotics. Thus antibiotic therapy may enhance acquisition of antibiotic resistance and other virulence factors.

Molecular tools make it possible to better trace the origin of infection in hospital settings and estimate the relative contribution of each source to ongoing transmission. It is technically feasible although quite challenging to directly estimate duration of colonization and transmission probabilities, which are critical parameters for planning effective interventions. The challenge is in the design and implementation of the study rather than in detection. To examine the potentially modifying effects of microbiota on transmission or identify predictors of horizontal gene exchange in vivo will require a much better characterization of the dynamics of microbiota and background rates of horizontal gene exchange. In addition to the study design and measurement challenges, the analysis will be quite complex. The complexities might be somewhat unraveled by applying ecological and evolutionary theory and using a system approach.

## 13.4 INTERACTIONS

One of the most exciting research areas made feasible by modern molecular tools is the potential to study the interactions among the microbes that live on the human host with each other and the human host, and how these interactions prevent or enhance spread of known human pathogens. At this writing, work on this area is beginning, and has been targeted by the U.S. National Institutes of Health. The Human Microbiome Project, as it is known, is currently in descriptive stages. (An introduction to the Human Microbiome Project is presented in Chapter 7.) We know very little about the microbes that live in and on the human body. In particular, we have limited understanding of the structure of the microbial communities, how they change with time, what their boundaries are, what possible communication they may have with the human immune system, and what their responses are to treatment and invasion by pathogens.

What we do know is that sanitation and medical intervention have profoundly influenced the human microbiota. For example, until recently, probably all humans carried eukaryotic parasites, such as worms, in their bowel. Undoubtedly, humans and their parasites coevolved, and the human immune system adapted to their presence. In the absence of these parasites, individuals highly adapted to their presence may develop disease. The absence of worms, which depress immune response, is hypothesized as one explanation for the epidemic of asthma observed throughout the United States. Other microbes adapted to the human host are becoming extinct, such as *Helicobacter pylori* (another microbe whose absence is associated with increases in asthma).[8] Additional microbes that coevolved with humans have undoubtedly become more rare or even extinct, not only because of changes in sanitation but also because of the use of antibiotics and other medical treatments. Treating bacterial infections with broad-spectrum antibiotics can be likened to using a nuclear bomb; there is significant collateral damage, because an antibiotic effective against a pathogen will be effective against other bacteria with similar characteristics. Cephalosporins are effective against *Streptococcus pneumoniae* and *S. aureus*, which are gram-positive bacteria, and also against gram-negative bacteria like *Escherichia coli* and other enteric bacteria (bacteria living in the intestinal tract). Therefore using cephalosporin to treat a case of pneumonia caused by *S. pneumoniae* also affects the bacteria found in the intestinal tract. The effects may be transient or longer term. In some cases, the microbiota does not return to a healthy state and is overgrown with organisms that cause chronic disease. This is the case with antibiotic-associated diarrhea caused by *Clostridium difficile*. *C. difficile* is a spore-forming microbe that produces a toxin that causes diarrhea. It was first identified as a cause of diarrhea among patients treated with the antibiotic clindamycin in 1978.[9] The diarrhea can be quite severe, and the bowel can be injured to the degree that surgery is necessary. Further, the spores are widely disseminated during defecation and very difficult to kill. *C. difficile*, which was relatively rare 30 years ago, is endemic in many hospitals and carried asymptomatically among a (presumably) larger portion of the general population. These individuals are at risk of severe diarrhea when treated with an antibiotic. Understanding the emergence and spread of *C. difficile* requires considering the bowel microbiota as a community, rather than focusing on a single pathogen. Considering the community as well as the individual pathogen is a paradigm shift for infectious disease epidemiology.

Epidemiologists have tended to consider the effects of each microbe in isolation, building on the seminal work by Robert Koch. The Koch–Henle postulates lay out a strategy for identifying a single microbe responsible for a specific disease. As even Koch recognized, the introduction of a known pathogen is often insufficient to cause disease. In addition to variations in host susceptibility due to prior exposure leading to immunity, intrinsic factors such as genetic variation in surface receptors or immune response, and factors that increase susceptibility such as malnutrition or injury, the presence of other microbes can either enhance or inhibit the ability of a pathogen to invade. Molecular techniques make it easier to study how microbes work together to cause (or prevent) disease in humans, because we are no longer constrained by the limits of microbiology based on culture. With culture, the growth conditions are designed to identify microbes we know how to grow. For example, the growth media contains nutrients and inhibiters, such as antibiotics, to select for the microbe of interest, and the culture is grown at specific temperatures and in the presence or absence of oxygen to optimize growth. These conditions were identified in painstaking studies over decades. With culture, the amount of the microbe of interest present can be determined, quantitatively or qualitatively depending on the sample, but how it compares to the total microbial load cannot be determined because the conditions are specific to the microbe of interest; the polymerase chain reaction (PCR) has changed all of that. PCR makes it possible to estimate the total bacterial load by using primers for highly conserved regions found in all bacteria, the ribosome. PCR can also be used to identify virus particles, fungi, eukaryotes, and archaea. With these tools and appropriate study design and analysis, it becomes feasible to examine how microbes might work together. (Note that PCR of DNA is not a panacea, as it cannot distinguish between dead and living microbes without special processing, and there are biases in detection.)

There is good evidence that there are synergies in transmission and pathogenesis among microbes. Genital ulcerative diseases enhance transmission of HIV, and HIV depresses immune function, which increases severity of some microbial infections, enabling growth of other microbes that usually do not cause disease in humans. Syphilis, herpes, and chancroid are sexually transmitted diseases that cause an open sore known as an ulcer. The ulcer is a door through which HIV can both enter and exit; persons with genital ulcers are at higher risk of acquiring and transmitting HIV. *Candida albicans*, a yeast, is a common commensal that usually does not invade tissue. In the presence of HIV, which suppresses host immune response, *C. albicans* invades the mouth mucosa causing a condition called thrush. Other viral infections predispose the host to bacterial infection. For example, risk of bacterial pneumonia increases following influenza infection, and risk of middle ear infections increases following many viral infections. In addition to two or more pathogens working together to enhance transmission or pathogenesis, the microbiota found on and in the human host can either enhance or inhibit transmission and pathogenesis. Molecular epidemiologic studies can greatly enhance our understanding of the underlying mechanisms behind these observations. The increase in knowledge will help identify new strategies for protecting human health.

At this writing, we have just begun to describe the microbiota; the few studies are limited in sample size and study population. Nonetheless the results have been surprising. The variation in microbiota between individuals and within an individual over time is much, much larger than previously imagined.[10,11] We have also learned that although the number of phyla that can live in and on the human host is large, the specific genera and species are constrained to those that can live within the temperature and nutrient constraints of the human body. The extent that differences in microbiota result in variations in health and disease remains to be explored, and is the subject of ongoing investigations. In addition to cataloging species present, some groups are conducting metagenomic studies, which catalog the genes present, based on the theory that there is redundancy in function across species present. Others are measuring outputs, studies of the transcripts, proteins, and metabolites. If these products vary predictably with health and disease, they are potential diagnostic and prognostic indicators.

Some of the differences observed in microbiota are probably attributable to differences between humans, because we all are unique in our genes, epigenome, diet, behaviors, and medical history. These differences must result in variations in our microbiota and our response to microbial infection. Individuals respond differently to the same microbe even when given the same dose. Some of these differences may be attributed to prior exposure, either natural or from a vaccine, which resulted in immunity, but there are also intrinsic differences. Mortality due to infection is more highly correlated between monozygotic twins than fraternal twins, and between parents and their biological offspring than their adopted offspring, which suggests that there is genetic variation in susceptibility.[12] Identifying genetic variations that predispose to specific infections can be an important lever for exploring mechanisms of pathogenesis, and identifying better diagnostic and treatment strategies.

While much can be learned about microbiota using model systems in the laboratory, there is an important role for epidemiologic studies in identifying the frequency that phenomena occur and their association with health and disease in vivo. Epidemiologists can take advantage of natural experiments that will enable us to observe what changes are wrought on our microbiota with medical intervention or over the course of disease. Surprisingly, our knowledge of microbial growth conditions in vivo is limited. Understanding these requirements can greatly assist in identifying optimal treatment strategies that aid the host without also assisting the pathogen. Molecular techniques make it increasingly feasible to identify these conditions, although the feasibility depends on the microbe and where it grows. There is some controversy, for example, over whether iron supplementation in the chronically ill increases risk of bacterial infection. Humans use iron to make hemoglobin, which carries oxygen. Iron is an essential element for bacteria too; thus, human response to infection is to sequester iron. The evidence that iron supplementation increases risk of bacterial infection is inconclusive, making it a fruitful area for research that might be expanded to examine the potential risks and benefits of other micronutrients on bacterial and viral growth in vivo.

Our understanding of disease pathogenesis also is based heavily on laboratory experiments. Even with current techniques, it is often difficult to detect whether putative virulence factors are expressed *in vivo*, or at what stage of transmission or pathogenesis a particular factor is expressed. Apparently virulent microbes can be isolated from disease-free individuals, sometimes without any prior disease manifestation. Much future work is needed to understand the determinants of virulence expression and the role of other microbes in mediating the expression or interaction of the host. Monitoring in vivo expression is also essential for designing vaccines. For some microbes, the goal of vaccination is sterilizing immunity, that is, to prevent colonization altogether. Sterilizing immunity is necessary if the long-term goal is disease eradication. Eradication was the strategy chosen for smallpox, polio, and measles. Note these are all viruses that are found solely in humans. Bacteria are more complex; nonpathogenic variants may predominate, and some may have a neutral or even positive impact on human health. Sterilizing immunity may not be desirable, as it may lead to colonization by potentially more pathogenic organisms. In this case, the appropriate strategy might be to vaccinate against specific virulence factors, exerting evolutionary pressure against their expression. The vaccine for diphtheria is not against the surface proteins of *Corynebacterium diphtheriae*, but against the toxin it sometimes expresses, which causes disease. In this case, vaccination exerts selective pressure against toxin production. A similar strategy might be taken for other pathogens, if the appropriate virulence factors can be identified.

A better understanding of the interactions among microbes and between microbes and the human host also has the potential to enhance diagnostic and prognostic indicators. As emphasized elsewhere in this book, but worth repeating here, it is highly likely that there are multiple species of microbes growing in what previously were thought to be monocultures. Focusing on a single pathogen misses the role that other organisms present might play in transmission or pathogenesis. It also misses information of diagnostic or prognostic value that monitoring co-colonizing (or co-infecting) organisms might provide. Microbial interactions can be positive, negative, or neutral. If there is a positive interaction, knowing that an additional microbe increases potential for virulence might modify treatment. The patient might be monitored more frequently if co-colonizing microbes have a neutral or negative interaction than a positive interaction with the pathogen.

There are multiple technical and analytical challenges to measuring and interpreting these interactions, which take place at multiple scales. Microbes interact with each other and with the host, and microbes are exchanged between hosts and between populations. The number of data points is enormous, and there is a potential for error in the measurement of each. Timing of measurement may be important, because what we think of as the pathogen may only be able to grow following growth of another microbe. We have yet to identify valid and reliable indicators of microbial community structures that might be associated with health and disease or propensity for transmission. Even if these indicators were available, theoretical and analytical frameworks are required for appropriate design, analysis, and interpretation.

## 13.5 CLOSING THOUGHTS

Molecular tools let us see the world anew. Increased discrimination enables the resolution of previously unsolvable problems while revealing previously hidden ones. The great insight of epidemiology is that by examining factors from the molecular to the individual to the human community we can identify factors that contribute to human health and disease. Integrating molecular biology with epidemiology multiplies the power of each field, an increase in power that will be needed as we face future challenges to human health. Climate change has already resulted in new patterns of disease, globalization has increased the rapidity of disease transmission, and urbanization has done both. Fifty years ago the collective wisdom was that we had beaten infectious diseases. Back then, the vision was some day we would be microbe free; Isaac Asimov described space travelers as having all microbes purged from their system to avoid contaminating space.[13] Now we know better. Microbes are *incredibly* resilient and adaptable and also are essential to our health. I hope readers will use the material presented in this text as a foundation to apply molecular tools to the myriad of health problems facing them, and to provide important insights into our understanding of disease transmission, pathogenesis, and evolution that can be translated into effective disease prevention strategies.

# References

1. S. Li, J.N. Eisenberg, I.H. Spicknall, et al., Dynamics and control of infections transmitted from person to person through the environment, Am. J. Epidemiol. 170 (2) (2009) 257–265.

2. M.J. Keeling, P. Rohani, Modeling Infectious Diseases in Humans and Animals, Princeton University Press, Princeton, 2008.

3. C.D. Heaney, E. Sams, S. Wing, et al., Contact with beach sand among beachgoers and risk of illness, Am. J. Epidemiol. 170 (2) (2009) 164–172.

4. I. Haase, S. Olson, M.A. Behr, et al., Use of geographic and genotyping tools to characterise tuberculosis transmission in Montreal, Int. J. Tuberc. Lung Dis. 11 (6) (2007) 632–638.

5. M.E. Kolader, N.H. Dukers, A.K. van der Bij, et al., Molecular epidemiology of *Neisseria gonorrhoeae* in Amsterdam, The Netherlands, shows distinct heterosexual and homosexual networks, J. Clin. Microbiol. 44 (8) (2006) 2689–2697.

6. P.K. Moonan, J. Oppong, B. Sahbazian, et al., What is the outcome of targeted tuberculosis screening based on universal genotyping and location? Am. J. Respir. Crit. Care Med. 174 (5) (2006) 599–604.

7. N. Wolfe, Preventing the next pandemic, Sci. Am. 300 (4) (2009) 76–81.

8. M.J. Blaser, S. Falkow, What are the consequences of the disappearing human microbiota? Nat. Rev. Microbiol. 7 (12) (2009) 887–894.

9. D.N. Gerding, *Clostridium difficile* 30 years on: what has, or has not, changed and why? Int. J. Antimicrob. Agents 33 (Suppl. 1) (2009) S2–8.

10. P.J. Turnbaugh, M. Hamady, T. Yatsunenko, et al., A core gut microbiome in obese and lean twins, Nature 457 (7228) (2009) 480–484.

11. E.A. Grice, H.H. Kong, S. Conlan, et al., Topographical and temporal diversity of the human skin microbiome, Science 324 (5931) (2009) 1190–1192.

12. M. Osler, L. Petersen, E. Prescott, et al., Genetic and environmental influences on the relation between parental social class and mortality, Int. J. Epidemiol. 35 (5) (2006) 1272–1277.

13. I. Asimov, The Robots of Dawn, Doubleday, Garden City, 1983.

Printed and bound by CPI Group (UK) Ltd, Croydon, CR0 4YY

03/10/2024

01040313-0020